Advance Praise i

"You Are Not Expected to Understand This"

"In truth, *'You Are Not Expected to Understand This'* is startlingly understandable! These vivid, lucid, brilliant essays tell the origin stories of coding, the secret infrastructure that shapes our online life. We meet the people who wrote and rewrote the lines of code that changed the world. We glimpse their ambitions, mistakes, remorse, fixes, and ingenuity. We understand why (and how) women were the ones who designed early programming languages like COBOL; how pop-up ads came to exist; how the 'like' button blew up news and politics as we knew them. Read this book, and you will never look at your newsfeed the same way again."

—**Liza Mundy,** author of *Code Girls: The Untold Story of the American Women Code Breakers of World War II*

"Code powers much of modern life, yet most of us spend little time thinking about it. This book will change that. Wide-ranging, provocative, and bursting with humanity, *'You Are Not Expected to Understand This'* is essential reading on the history and culture of code."

—**Sara Wachter-Boettcher,** author of *Technically Wrong: Sexist Apps, Biased Algorithms, and Other Threats of Toxic Tech*

"Code governs our lives—and this book does a delightful job of giving us a glimpse into some of the biggest wins, and most colossal blunders, in software."

—**Clive Thompson,** author of *Coders: The Making of a New Tribe and the Remaking of the World*

“You Are Not Expected to Understand This”

"You Are Not Expected to Understand This"

How 26 Lines of Code Changed the World

Edited by Torie Bosch

With an introduction by Ellen Ullman
and illustrations by Kelly Chudler

Princeton University Press / Princeton & Oxford

Published by Princeton University Press
41 William Street, Princeton, New Jersey 08540
99 Banbury Road, Oxford OX2 6JX

press.princeton.edu

Library of Congress Cataloging-in-Publication Data

Names: Bosch, Torie, editor. | Chudler, Kelly S., illustrator. | Ullman, Ellen, writer of introduction.
Title: You are not expected to understand this : how 26 lines of code changed the world / edited by Torie Bosch ; with an introduction by Ellen Ullman and illustrations by Kelly Chudler.
Description: First edition. | Princeton : Princeton University Press, [2022] | Includes bibliographical references and index.
Identifiers: LCCN 2022013091 (print) | LCCN 2022013092 (ebook) | ISBN 9780691208480 (pbk. ; acid-free paper) | ISBN 9780691230818 (e-book)
Subjects: LCSH: Computer programming—Popular works. | Computer science—Social aspects—Popular works. | BISAC: COMPUTERS / Programming / General | SOCIAL SCIENCE / Technology Studies
Classification: LCC QA76.6 .Y585 2022 (print) | LCC QA76.6 (ebook) | DDC 005.13—dc23/eng/20220527
LC record available at https://lccn.loc.gov/2022013091
LC ebook record available at https://lccn.loc.gov/2022013092

British Library Cataloging-in-Publication Data is available

Editorial: Hallie Stebbins, Kristen Hop, and Kiran Pandey
Production Editorial: Natalie Baan
Text and Cover Design: Chris Ferrante
Production: Danielle Amatucci and Lauren Reese
Publicity: Kate Farquhar-Thomson and Sara Henning-Stout
Copyeditor: Michele Rosen

Page 132: Comic adapted from MonkeyUser, reproduced with permission.

This book has been composed in IBM Plex

Printed on acid-free paper. ∞

Printed in the United States of America

10 9 8 7 6 5 4 3 2 1

Contents

Preface

Torie Bosch

In high school in the late '90s, I took my first and only coding class—a course on C++. I encountered all of the problems that you hear about when it comes to girls and programming: the only girl in the class, I was coming to the subject cold. Though the class technically had no prerequisites, every other student had a working understanding of coding; many of them had learned from their fathers. My teacher was a woman and started out encouraging, but quickly became exasperated with me. From the first assignment, which I believe had to do with a string of numbers, I flailed. I eked out an A in the class, helped by the boys, who would spot problems in my code and tell me how to fix them. But I never really understood what I was doing—how programming *worked* or what the different languages meant. The teacher largely skipped over that stuff because she assumed we all knew it, and, I assume, she didn't want to hold the rest of the class back for me. I've always wondered if she was frustrated with me for making women and girls in STEM look bad.

Before going into that C++ class, I had thought of programming as something simple and straightforward: you tell a computer what to do, and it executes. But that class demonstrated, on a small and annoying scale, that "telling a computer what to do" inherently requires messy human thinking. I learned that code could be "wrong" yet still somehow work, and that it could have unintended ramifications—ramifications that might not matter much when I was working on, say, a birthdate calculator, but would matter a lot if I were working on a transportation system. Thinking about these issues was fascinating; coding was not for me, perhaps, but *thinking* about it was.

That's the idea behind this book: that we should all think a little more about code, because code has in infinite ways changed how we live in the world, for better, worse, or somewhere in between. And behind the code, of course, are people: people who make decisions, make mistakes, make assumptions, take brilliant chances, and take shortcuts, with major—and sometimes unintended—ramifications. The 26 essays in this book, written by technologists, historians, journalists, academics, and sometimes the coders themselves, tell the stories of people writing code that is by turns brilliant, funny, sloppy, and shortsighted. The essays show us how code works—or how, sometimes, it doesn't work—owing in no small way to the people behind it.

This book grew out of an article published by *Slate* in October 2019 titled "The Lines of Code That Changed Everything." From the time I began working on that project, I struggled with how to define "lines" of code. A line of code can be, literally, a single string of code—a line of BASIC, say. But code here also means the bigger ideas that underlie software programs, programming languages, digital platforms, and physical hardware, and the interaction of these elements is what defines our digital world today. In some cases, we can no more easily separate them than we can separate technology from the humans who created it. This book embraces the messy.

Many of the essays will focus on actual lines of code. For instance, James Grimmelmann writes about a T-shirt with four lines of Perl code—the RSA[1] encryption algorithm—that was written to protest US export controls around encryption. Thanks to that activism and the clever use of code, more people today have the ability to communicate securely. Josephine Wolff writes about Heartbleed, one of the most pernicious vulnerabilities in computing history, and the code mistake that made it possible. Ellen Stofan and Nick Partridge look at the Apollo 11 lunar module's code and an error alert that nearly ended the first moon landing.

But some of the essays embrace a broader definition of "lines of code." For instance, I've chosen to include essays that focus not on code per se, but on software. Charlton McIlwain, author of the landmark book *Black Software*, examines the first Police

Beat Algorithm, created in the 1970s, and how it led to today's predictive policing software. McIlwain's essay demonstrates how code cannot possibly be free of human biases around race. Afsaneh Rigot and Mahsa Alimardani write about the Telegram channels that first helped Iranians evade government surveillance and censorship, only to end up putting Iranians—especially members of the LGBTQ community—at risk.

Finally, this book includes essays on coding languages and on the act of coding itself. Claire L. Evans writes about the birth of COBOL and the much-overlooked role of women in early computing history. Meredith Broussard writes about how the gender binary and the code binary are intertwined. Ethan Zuckerman's essay looks at the role he played in creating a scourge of the Web: the pop-up ad. David Cassel examines how a perfectly anodyne comment in code—"`You are not expected to understand this`"—became part of programming culture and, as he put it, a "cherished reminder of a momentary glow of humanity in a world of unforgiving logic."

To narrowly focus on actual lines of code would have overly constrained the collection, overlooking many programming "mistakes," moments of genius, and human decisions and biases that are infused into society through thousands of lines of code. Indeed, beyond code, these essays have a second, no less important focus: people. The book tells the stories of the people behind the programming and the people affected by it—a group that includes all of us. To me, certainly, these stories all feel personal: around the same time Zuckerman was working on the first pop-up ad for Tripod, I was building my own site on Tripod (devoted to, of all things, Backstreet Boys fan fiction), which would come to host its fair share of pop-up ads. It was also the site that taught me how to use HTML: the HTML links, as Brian McCollough writes in his essay, that were and are foundational to the Internet.

These essays should feel personal to all of us, coders or not, because the technological successes and failures and strange moments they describe undergird our lives. The decisions made by the people in this book shape our online behavior: how we learn, how we interact, how we define ourselves. And in the twenty-first century, our online lives bleed into our offline lives, to the point where sometimes we can't distinguish between them.

I became interested in the Internet—and C++—thanks to my mother, who bought my family a computer and signed up for AOL back in 1995. She had a tech background herself, having learned programming in the 1970s when she was a consultant. She never particularly enjoyed it, though, in part because she felt her questions were often dismissed—especially if they concerned the bigger picture, beyond the work at hand. Just before Y2K, she told me that she once asked an instructor why programmers used two digits to represent years instead of four. "Won't that cause problems?" she asked him. "Someone else will have to deal with it," he told her. Programming mistakes, quick shortcuts that become permanent fixtures, flashes of brilliance—all of us have to deal with them.

Introduction
Error, Failure, and Code Creation

Ellen Ullman

You need the willingness to fail all the time.

Those words guided me throughout all the years when I worked to become a decent programmer, as they no doubt guided countless others. That one sentence reminded us that coding is a life in which failure will be your constant shadow. Bugs, crashes, halts, glitches, hacks: programmers who want to survive in the profession (like anyone hoping to create a new thing on earth) must come to a begrudging acceptance of failure as a confounding helper, an agent of destruction you wish you could evade, but never can.

The words were spoken by John Backus, who led the group that created the FORTRAN programming language, fully released in 1957.[1] FORTRAN (short for Formula Translator) was the first language that allowed programmers to write code that was not directly tied to any one computing environment. It was a frustrating project that lurched from failure to failure. Backus went on to say:

> You have to generate many ideas and then you have to work very hard only to discover that they don't work. And you keep doing that over and over until you find one that does work.[2]

He also told us:

> If you are not failing a lot, you are probably not being as creative as you could be—you aren't stretching your imagination.[3]

Software companies try to avoid serious failures with procedures, rules, reviews. But programs are works of the imagination that must then make the hazardous crossing into the structured world of code. The attempts to avoid failure will also fail.

All code has flaws, inevitably. Human thought is wonderfully chaotic; it allows us to hold incompatible beliefs, be interrupted, function in a world we do not fully understand. So much of what we know is inscribed in the body, the product of evolution, instinctive, not readily accessible to the rational mind, what Daniel Kahneman has described as fast thinking (System 1). Meanwhile, code-writing (as opposed to the creative work of code-design) requires fully conscious and rational thought, Kahneman's "slow thinking" (System 2),[4] a level of focused attention that is impossible to sustain over time.

I have a friend who was once in charge of testing at a startup that was frantic to go public. The IPO was delayed for months on end because of the relentless appearance of new serious bugs. The higher-ups demanded to know, "When will all the bugs be found?" It was a ridiculous question, because the testing was being done even while new code was being written. Meanwhile, "fixes" to already discovered bugs were in the business of creating a whole new pile of bugs. In any case, no one can predict when the last bugs will be found, because the only correct answer is, "Never."

Many bugs are blind spots in the code. The designer and programmer try to protect the system by looking for conditions that will break things: they will not find them all. Most often, software teams are rushed. They have to create systems quickly. Programmers don't have time to lean back, think of other things, let the background of the mind speak. A prime source of bugs is absurd scheduling.

Other bugs are like physical vulnerabilities inscribed in the DNA. These bugs sit quietly until some environmental factor (in humans, things like ageing, chemicals, medications) suddenly activates the flaw, and we get sick. In the case of computing, the technical environment is complex and constantly changing. Programs interact with modules not foreseen in the original design; with new operating systems and changed ones, variations in chips, network configurations, protocols, device drivers; bedeviled by documentation that cannot keep up with the changes. What worked one day doesn't work the next, and the programmer's constant question is, "What changed?" Well, lots of things. Which one (or ones) did the damage? That way lies madness.

The deepest weaknesses are revealed when a digital creation is designed for expert users in a collegial environment, and then opened to a wider pool.

Dennis Ritchie and his team developed the C language,[5] which, along with Unix, was part of a research project conducted inside the storied Bell Labs technology incubator.[6] The language gave the team's programmers a great deal of freedom, including direct access to the contents of memory, something systems normally do not allow, in order to protect the integrity of the coding environment. That level of freedom was fine as long as their work remained a research project. According to Brian Kernighan, who coauthored the book that introduced C to the world,[7] Ritchie did not anticipate that the operating system and language would become "as big as they did."[8] Yet they did indeed become big. Programmers' access to memory then escaped into the wild: programs acquired the dangerous ability to invade and manipulate the memory space of another program (mostly by accident), and that invaded program can invade another's (and so on), enabling a world of perplexing bugs.

Then there is the Internet itself, derived from the ARPANET, which was created as a platform in which a limited group of researchers could converse openly about scientific subjects.[9] Security was not assumed to be needed. And so arrived the hackable digital universe.

I once had the good fortune of working for a hacker. This goes back to the time when "hacker" was an honorific, as it still is among talented hardware and software engineers. It refers to a type of crusty programmer who can chop through code with a combination of grim determination and giddy enthusiasm. The goal, above all, is to uncover the flaws that induce the failures, then (somehow or other) devise the fix that will make things *work*. Their solutions are often "ugly," in coder parlance (aka kludges), the product of down-and-dirty plumbing. But no matter. Maybe lovely, elegant programs and systems can come later. Or not.

"Hacker" has acquired a less admirable meaning, of course, having acquired the taint of what we used to call "crackers," as in safe crackers, people not allowed to get at what's in the safe but who get in anyway. It is a chaotic world involving everyone from cryptocurrency tinkerers to bank thieves; from hackers working for hostile nation states to ones stealing data for espionage and ransom; to those seen as ethical hackers, who want to reveal the wrongdoings of anyone or anything in power; to loners looking for notoriety; to pranksters, jokers, naughty boys of all ages, breaking in just to see if they are clever enough to do it. (It's *fun* to make porn appear in Zoom meetings, isn't it?)

There are the workaday hacks, the constant reports of code vulnerabilities. Peter G. Neumann, the revered computer science researcher, moderates "The Risks Digest,"[10] which is updated weekly, sometimes as often as every few days. The "Crypto-Gram Newsletter,"[11] written by noted security analyst Bruce Schneier, is released monthly. As individual programmers and software makers struggle against the onslaught of flaws in their own code, they are meanwhile bombarded by the hacks that rain down upon the digital planet, nearly invisible, like the solar wind.

Then come the hackers who break into the code meant to defend against hackers: code to protect code becomes a victim. NASA stored reports of vulnerabilities they received from friendly hackers, and then the store *itself* was hacked.[12] Software written by the company CodeCov,[13] which is widely used to test for bugs and code vulnerabilities, was broken into by Russian hackers, giving them a window into the very code to be protected. In a recently revealed 10-year-old hack, Chinese spies broke into RSA's cryptosystem.[14] The company is a corporate security giant whose customers include "tens of millions of users in government and military agencies, defense contractors, and countless corporations around the world," according to wired.com. The break-in allowed "hackers to instantly bypass [RSA's] security system anywhere in the world."[15]

The fate of humanity hangs in the balance. Nicole Perlroth's book *This Is How They Tell Me the World Ends: The Cyberweapons Arms Race*,[16] describes how the Internet—buggy and hackable—has become a potent military tool. It has the dark power to ignite global war: by accident, or by design.

Now I will return to the "good" use of hacker, because I want to preserve its historical meaning among the general public and give the original hackers their due: an army of sometimes disheveled geniuses who were wary of rules and formalities, nonconformist in their thinking, somehow both brilliant and practical at the same time, who could reach in, rummage around, and figure out what to do. A member of Backus's FORTRAN team called their group "the hackers of those days."[17]

A now-famous hack saved the Apollo 13 mission from disaster.[18] Before the mission could achieve a moon landing as planned, an oxygen tank exploded in the command module. The three astronauts had to take refuge in the lunar module, which was designed to carry only two astronauts. To reduce the build-up of carbon dioxide, they retrieved an additional canister of lithium hydroxide pellets (a carbon dioxide scrubber) from the command module.[19] But there arose the sort of problem that plagues complex projects: components designed and built separately. One canister had a round connector, the other a square one, the proverbial square peg in a round hole.[20] A remedy had to be found—quickly—or all three men would die of asphyxiation.

NASA engineers on the ground raced to find a solution. They threw together bits of stuff that were on the spacecraft—plastic bags, covers ripped from manuals, duct tape, cardboard, anything—and devised a bridge between the mismatched connectors. It was one of those "ugly" fixes. As the Apollo 13 astronaut James Lovell later described it: "Hose. Duct tape and an old sock."[21]

Daniel Kaminsky, a famed cybersecurity expert, created another legendary, down-and-dirty hack. In 2008,[22] he discovered a security hole in the Internet's Domain Name System (DNS), which converts website URLs to specific IP addresses. Kaminsky saw how easy it was for knowledgeable bad actors to redirect the user not to the intended destination, but to a world of fake sites—a "bank," a "credit card company," an "email login"—and therefore collect the user's IDs and passwords. He alerted others and, along with Paul Vixie, coded an emergency patch.

Kaminsky, who will forever have a place of honor among the greats of the hacker community, died on April 23, 2021. His obituary in the *New York Times* called him "the savior of the Internet."[23] He was the first to sound the alarm and respond to the threat. Yet, given what we know about the relationship between coding and error, it is no surprise to learn that the patch was far from perfect. After the "fix" was installed, there were 6:00 a.m. calls from Finnish certificate authorities saying their security procedures were broken. Some DNS servers stopped working correctly. And there were some pretty harsh words from Kaminsky's peers in the security community.[24] Years later, in a talk at the 2016 Black Hat hacker conference, Kaminsky referred to his patch as "that DNS mess."[25] Vixie, a longtime steward of the DNS, described the code they cobbled together in terms yet more ugly than Apollo's old sock: he compared it to dog excrement. In the way of hacker expediency, he called it the best dog excrement "we could have come up with."[26]

Each of the programs, systems, and concepts discussed in this book had to go through the test of trial-by-error. The essays in this book explore a wide range of topics. Several offer a deeper look at technologies familiar to the general public: the coming of Email, hyperlinking, JPEG image files, the Facebook Like. Some discuss historical landmarks that ought to be known more widely: women's contributions to early computing; the creation and endurance of COBOL, the first language in general use for business software; the coming of BASIC, the wonderful beginner's language.

Two essays explore deeper concepts in computing: data encryption, and the Markov Chain Monte-Carlo concept (MCMC), a foundational mathematical method used to understand distributions in data and arrive at probabilities.

Computing can bring happiness, as three essays show. There is pleasure (and pain) in learning to write code; in the fun brought into the world by Spacewar!, the first distributed video game; and in the advent of the Roomba, which, in addition to cleaning floors, also gave hours of delirious pleasure to innumerable cats.

Two essays discuss contributions to computing that I see as being derived from the idea of "the wisdom of the crowd": the Facebook Like button and page ranking. The premise is that numbers in and of themselves say something about the worth of whatever is being liked, from websites to Instagram postings to dance crazes on TikTok: more likes equals more eyeballs equals "better." The underlying theory is based on the belief that, given a very large universe of participants, a truth will emerge.

The coming of the "smart mob" has been a decidedly mixed blessing. Twenty-five years ago, I had an informal talk with Larry Page about Google's search engine as it worked at the time. I said I was concerned that the order in which results were listed, based as it was on the number of links into a given page, was a species of the rich getting richer. Larry, ever thoughtful, sat quietly, considering his reply. Finally he said, "I worried about that too, but I realized there was nothing I could do about it."

What he meant was that there was nothing he could do *algorithmically*. Given the immense universe of knowledge, a human curator would have faced an impossible task; code has to be the curator. Google's search engine has improved vastly over time, its criteria for ranking becoming ever more sophisticated. And search engines, most modeled on Google's, have brought astounding advances in how human beings can understand the world. Yet search engines have also ushered in the age of "most popular," "trending," "bests," and posts that users hope will "go viral." This amplification of responses can empower the public and create a world of fun. They also reveal the hazards of assigning wisdom to the crowd: results prejudiced by the cultural majority, an arms race between the search algorithm and sites wanting to promote themselves, conspiracy theories, hordes of influencers stoking likes and clicks, truly fake news.

Then there are the programs we wish had not survived the assault by bugs. One essay examines so-called predictive policing, which pretends to predict where crime will take place in the future. Like all AI algorithms, it is based on databases laced with bad information, on methods that are rife with bias.

On a lighter note, there is another maybe-we-never-wished-for code invention: the pop-up ad. The essay here, by the programmer who authored it, describes his remorse, the regret he feels about loosing the pop-up upon the world.

A book about code must necessarily address the subjects that are integral to the creation of software: error and failure. "The Lost Mars Climate Orbiter" describes a failure that, 28 years after Apollo 13,[27] echoes the earlier mission's mistake: system parts created separately. One team used the American measurement system, the other the English Imperial system. The repetition of this type of error shows how pervasive are the hazards in complex systems, where one group of engineers cannot possibly create the whole, and disparate parts must somehow be knit together, and flawlessly.

"Heartbleed" describes a bug deep in the internals of the Internet that caused havoc for millions of devices. A hacker exploited weaknesses in open-source software and vulnerabilities in the C language, as mentioned above, which gave programmers direct access to the contents of memory. Like so many errors, the problem lay dormant, everything apparently working, until something in the environment changed: the arrival of a hacker with malicious intent.

Another essay discusses the Morris Worm, the first to be distributed via the Internet. Robert Tappan Morris, then a graduate student at Cornell, wrote the invasive code as an intellectual project, as a test of the Internet's weaknesses. However, a mistake in his code instructed the worm to keep reproducing itself, whether or not a system had already been infected. Then he inadvertently released the worm into the wild. A senior engineer who worked on the emergency caused by the worm, Colm MacCárthaigh, later said, "It felt like the Internet was on fire." Morris never intended to cause the vast damage he did. In this sense, his worm was a bug inside a hack.

A particularly pernicious use of errant code was deployed by Volkswagen to falsely lower the readings of pollution levels caused by their diesel engines: an intentional bug, an error created for corporate gain.

And then we come to the day-to-day, unglamorous but vital chore performed by all good programmers: adding comments to their code. Comments are an invaluable tool; they describe sections of the program that are tricky, not immediately obvious or readable. Comments are acts of generosity, help for the unknown colleagues who will work on the code over time, in the hope that they will keep a system working.

Sometimes the "future" programmer will be the original author of the code, and the comment is a gift to oneself, since it is all but impossible for individuals to recall all the complex details in the software they have written. A bug is an opportunist that waits at the gate of any change to the body of running code; a comment is a weapon that, *a priori*, takes up the battle against software entropy.

I am just old enough to remember the desperate attempts by the United States to match the Soviet Union's great achievement, Sputnik, the first earth-orbiting satellite. NASA's launches were broadcast on television, some live. We saw one rocket after another exploding spectacularly on the pad; or collapsing in a ball of fire after lifting-off a mere few feet; or managing to rise into the sky only to burst into flames at the first stage of separation.[28] Those failures are engraved in the memories of those who watched the attempts: the great anguish inherent in technological achievement, and, per Backus, the imperative to try again.

Decades later, after scores of intervening successes—including a human's trip to the moon and projects that sent explorer satellites to the edge of our solar system and beyond—NASA launched the mission to send the Perseverance Rover to Mars. The launch took place on July 30, 2020.[29] On February 18, 2021, nearly six months later, Perseverance landed on Mars.

The landing was streamed live[30] thanks to NASA's commitment to inform the public, even if a mission might fail. What riveted my attention was a pane on the left side of the screen. It highlighted each stage as the mission unfolded, modules for launch, separations, cruise balance, etc. Between each module was a step that began with the word "`Interface`," as in: `Module A`, `Interface to module B`, `Module B`, `Interface to Module C`, `Module C`, and so on. You could see the tension in the faces of the women and men staring into their monitoring screens. I held my breath along with them.

There is no more hazardous place in a complex project than the handshake between one section and the next. In this interregnum lurks all the potential misunderstandings between separate groups of developers, as we saw with the lost Mars orbiter and the near catastrophe of Apollo 13. The illuminated word "`Interface`" always seemed to linger for far too long. I wondered if this latest generation had learned the lessons of their forebears, who knew the danger zones. In the case of a breakdown, did these young engineers have the hackers' skills to scrounge around and repair a ripped seam? This Mars Rover project seemed impossibly complicated, riddled with opportunities for disaster. I watched in a mood of both exaltation and horror.

Time went by. The display followed the steps in the project: one module, interface, next module, interface, and the next. Finally we came to the astounding unfurling of the parachute that gently lowered Perseverance to the surface. And it was done.

And yet.

There is no such thing as the last bug.

The problem appeared in the initial test of the small helicopter, Ingenuity, which had arrived on Mars attached to the underbelly of Perseverance, like a baby kangaroo in the pouch of the mother ship. Ingenuity was to attempt to fly in the thin atmosphere of Mars, to pioneer an age of powered, controlled flight—engineered by humans—on a planet other than earth.

The first try failed. The helicopter's start-up activities took longer than expected, and its computer shut down the motors. The engineers overseeing the mission identified a potential workaround and devised a patch. Yet, knowing that touching existing code is an excellent opportunity to break it, they wisely did not install it. Instead, they adjusted the commands they would send to the craft.[31]

Here was a repair that was sent not through the Internet but across 130 million miles of space.[32] Engineers had to wait two anxious earth days to find out if their changes would work.[33] On April 19, 2021, Ingenuity rose 10 feet into the Martian atmosphere as planned, hovered briefly, banked, turned, and landed at its takeoff point.[34]

More flights followed. Failure had led to success. This was a bug-fix for our time, another hack for the ages.

1

The First Line of Code

Elena Botella

What was the first line of code? It depends, a bit, on how exactly you define *code*.

For now, let's say *code* is a set of instructions that are given to a machine. Giving instructions to machines is something most of us do all the time. So far this morning, even before I opened my laptop, I flipped a light switch to turn on and off my bathroom lights, pushed a lever to instruct my toilet to flush, and pressed a button to tell my coffee grinder to grind some coffee beans. Each of these individual buttons or switches delivered a single instruction, "`on`" or "`off`," "do" or "do not," or, shall we say, "`1`" or "`0`," to machines that each knew how to do exactly one thing: emit light, flush water, or grind coffee beans.

Code happens when you have a machine that's ready to listen to not just one instruction, but very long sets of instructions, combining those "`on`" or "`offs`," "`1s`" or "`0s`," "dos" or "do nots" in practically infinite, seemingly magical combinations. And importantly, the difference between *code* and mere *language* is that code should always produce the same output given the same input. When I run lines of code, even on a different computer than the one I wrote them on, they should always produce the same result.

By this definition, the first code may have been written by Basile Bouchon in 1725, more than a century before the invention of the electrical generator.[1]

Basile Bouchon was a French weaver, at a time when patterned silks were the height of luxury and of fashion. Before Bouchon, it took several weeks of painstaking and tedious labor (tying knots in the loom strings) just to reconfigure a loom to the selected pattern.[2] As historian Abbot Payson Usher wrote in 1929, this "work was heavy, the hours long, and serious physical disorders were developed by this most wretched class of silk workers."[3]

Bouchon's solution? The first punch card, or, more specifically, a very long sheet of paper that would be "coded" by punching holes in the paper. This long sheet of paper was placed underneath a row of needles and gradually unspooled by the weaver. The holes (or lack of holes) in the perforated paper told the loom which needles to retract and which not to retract, which in turn changed the design of the fabric.

In practice, Bouchon's loom didn't work very well, and it was never widely used. But even at the time, other weavers could see that the idea held promise. It only took a few years for other people to start making improvements, notably by replacing the continuous roll of perforated paper with sturdier, more versatile punch cards. A few years after that, someone designed a loom that eliminated the need for a human to stand by to switch the punch cards, doing so mechanically instead.

Bouchon wasn't famous in his own era, but another man, Joseph Marie Jacquard, became rich and famous in the early 1800s for improving upon Bouchon's design, combining it with other innovations to create what became known as the "Jacquard Loom."[4] Emperor Napoleon Bonaparte was so impressed with Jacquard that, after personally visiting him, he awarded Jacquard an annual pension and a bonus for every Jacquard Loom manufactured in France.[5] Jacquard became a household name, but Bouchon was reintroduced to history books when British academic Reverend Robert Willis wrote that "the merit of Jacquard is not . . . that of an inventor, but of an experienced workman, who by combining together the best parts of the machines of his predecessors in the same line, succeeds for the first time in obtaining an arrangement sufficiently practical to be generally employed," giving the credit specifically to Bouchon for the use of perforated paper to program the loom.[6]

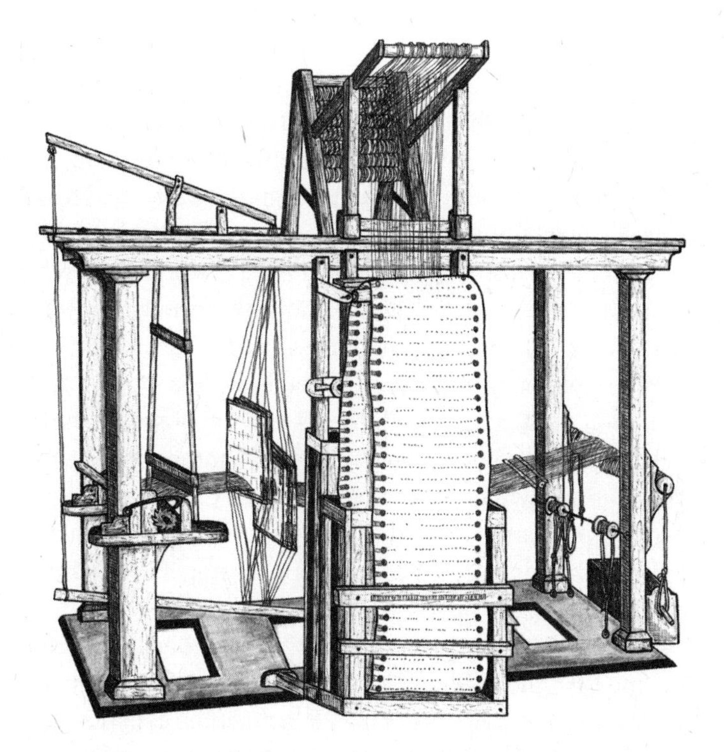

Basile Bouchon's loom, which was controlled by perforated paper tape, an early predecessor of the punch card.

The thread from Bouchon's first lines of code to the Jacquard Loom to the code that controls your computers and smartphone is clear and direct. Ada Lovelace and Charles Babbage, who laid out the blueprint for the first computer in 1837 (they called it the "Analytical Engine"), referenced the Jacquard Loom to explain how computers would work, and Charles Babbage even owned a silk portrait of Jacquard that had been coded on a Jacquard loom.[7] A full working model of Babbage and Lovelace's machine was never built—it would have been roughly the size and weight of a train locomotive, and powered by its own steam engine—but modern engineers believe that the Analytical Engine would have been quite powerful, albeit bulky, very slow, and without much storage capacity.[8] In fact, Alan Turing pointed out that

Charles Babbage's Analytical Engine could do the same set of essential tasks that any 1950s computer could complete.

In that paper, Turing explained what made a *computer* a computer: the ability to perform absolutely any computation. As long as your machine could do that, it didn't even need to use electricity, like Babbage and Lovelace's mechanical invention. The machine did need to have three things: a "store" or memory to keep track of data, an "executive unit" to carry out operations, and a "control" or set of instructions—in other words, *code*. Both the data in memory and the code could be described, Turing pointed out, as a set of numbers (0s and 1s), just like the binary of Bouchon's perforated paper roll. (In the same article, Turing defined what came to be known as the Turing Test or the "imitation game"—the idea that you could test the quality of an artificially intelligent computer by seeing if it could trick someone into believing it was a human.)

I said, at the beginning of this essay, that Basile Bouchon wrote the first lines of code if you accepted my definition: that code is a set of instructions, given to a machine capable of accepting long sets of instructions, that produces a consistent output. But some, including Turing, might say that definition is insufficient—or, as modern computer scientists would put it, Bouchon's machine and code weren't "Turing complete."

And why? Well, the loom wasn't a *computer*, in the sense that the loom or the punch cards couldn't *compute* anything. In his 1950 paper, Alan Turing used the term "universal machine" to describe any machine with the same capability as Lovelace and Babbage's invention: the capability to compute anything that it is possible to compute. "The existence of machines with this property," Turing wrote, "has the important consequence that, considerations of speed apart, it is unnecessary to design various new machines to do various computing processes."[9] In other words, since Bouchon's loom didn't compute anything, it wasn't a computer, so his first perforated paper tape wasn't really code. Computer scientists now use the term "Turing complete" to describe computers and programming languages that meet the standards of Turing's universal machine: virtually all computers and programming languages today qualify.

If you've written code, maybe in a computer science class, or maybe at work, it may not have always felt like you were *computing* numbers, especially if you just wrote a program to print "`Hello, World.`" But underneath it all, even that simple line of code, if executed on a computer, *did* involve a computation. Everything your computer or smartphone does boils down to a set of math problems: even displaying graphics is a matter of computation. Maybe, if you're a gamer, you've bought a graphics card (also called a GPU) for your computer, so your games would render more quickly—the difference between a GPU and a CPU is just the *types* of computations each kind of chip can execute more quickly. GPUs are best at complicated math problems with a lot of steps you can do simultaneously, since those are the types of math problems used to render graphics, while CPUs are best at complicated math problems whose steps need to be done sequentially. By the "Turing completeness" definition of a programming language, Bouchon's punch cards weren't code. By this definition, even HTML isn't code, since you can use HTML to design websites, but not to do most computations.

Basile Bouchon may have invented the use of "binary" to give long sets of instructions to a machine, but he didn't realize the invention's full revolutionary potential. In 1889, a German immigrant, Herman Hollerith, wrote his doctoral thesis at Columbia University explaining how to use punch cards to more quickly compute data. His system was used by the US Census Bureau to complete the 1890 census, cutting down the required amount of computation time by *at least six years*. The company that Hollerith founded, the Tabulating Machine Company, eventually became known as IBM.[10] Scholars debate the extent to which Hollerith borrowed directly from the Jacquard Loom, but IBM, for its part, does give Jacquard partial credit for inspiring Hollerith's design.[11]

Just as you could argue that Bouchon's machine instructions were too simple to be considered *code*, you could also argue the first lines of code were written before Bouchon. Other people "programmed" or "coded" machines before his birth, especially to make music. The Banu Musa brothers, who lived in Baghdad in the eighth century, developed an "automatic flute player" powered by a waterwheel, believed to be the first instrument capable of playing a preprogrammed melody, and probably the first programmable machine in general. Bouchon himself came from a musical background: his father was an organ maker.

It's probably no coincidence that Bouchon had exposure to the construction of musical instruments from a young age, but the connections among looms, musical instruments, and early computers run even deeper. Just as Herman Hollerith of IBM drew inspiration from weavers, he also probably learned from the player pianos that were commonplace in American homes during Hollerith's era. These player pianos used a perforated roll of paper to play a melody, not dissimilar from how perforated paper dictated the patterns on Bouchon's looms. At their peak in the early 1920s, player pianos, capable of performing automated tunes without human control, outnumbered conventional pianos by nearly 50 percent.[12] In a sense, the patterns of fabric and the verses of songs became the "loops" in modern computer code.

Code as *holes in paper* deserves a special spot in this book because of how important punch cards became as a way of representing both data and the code used to manipulate data. Binary, or the representation of data using 0s and 1s, is still the basic building block of all code and all computing: when computers became electronic, these 0s and 1s reflected the presence or absence of an electric signal.

The punch card, it turns out, is all you need to represent any number, any poem, any song, any computer program: practically anything at all.

2

Monte Carlo Algorithms

Random Numbers in Computing from the H-Bomb to Today

Benjamin Pope

The first code run on the first modern programmable computer remains a classified secret, but we know what it did: the ENIAC[1] in Philadelphia took 20 seconds to confirm to American scientists, only months after devastating Hiroshima and Nagasaki with nuclear fission weapons, that a much more powerful hydrogen bomb was possible. The superpowers' arms race demanded ever-more-powerful computers to calculate designs for weapons, and the algorithm of choice was the Monte Carlo method.

The US nuclear weapons program attracted the eccentric and talented to its base in Los Alamos, New Mexico, none more so than Edward Teller. Born into a Hungarian Jewish family and educated in Germany, he had escaped after the rise of Hitler and settled in the United States. He was so passionate about developing fusion weapons that he took pride in what Soviet propaganda called him: "the cannibal."[2]

His recruits to the nuclear program included a married couple: Marshall and Arianna Rosenbluth, who had both earned their PhDs at only 22 (at Chicago and Harvard, respectively) and met as postdoctoral fellows at Stanford. Unlike Teller, their son Alan remembers that they "thought of their weapons work as a kind of national service" but were politically liberal and would later become advocates for arms control.

Teller's colleague Stanislaw Ulam liked to gamble. While recovering from illness in 1946, he thought about card games:[3] How often will the house win in the notoriously hard Canfield solitaire? There are far too many possibilities to calculate for 52 cards—but Ulam realized he could get a fair approximation by shuffling the cards a hundred times and counting the rate of success.

This gave him and his colleague John von Neumann the solution to a more important problem: calculating how neutrons would flow in a nuclear warhead. You could follow a representative sample of simulated 100 neutrons as they collide with atoms, deciding using computer-generated random numbers what happens next: How far do they fly? Do they bounce off an atom, or are they absorbed and lost, or do they split the atom and form more neutrons? If these initial 100 increase in number on average, you have a runaway nuclear reaction. Using ENIAC, this calculation would be fast enough to design ever-more-powerful weapons. Their colleague Nick Metropolis suggested a name for using random numbers to guide a simulation: the Monte Carlo method, after the world's most famous casino, where Ulam's uncle used to gamble.

A central question in simulating the H-bomb was the *equation of state*, which describes how matter responds to changes in temperature and pressure. Marshall recalled working "60 hours or more per week trying to understand the physics of the interactions of radiation and matter under these extreme conditions."[4] They decided to tackle a related problem: the melting of a solid into a liquid. Directly computing the motions of hundreds of molecules, let alone trillions, would be impossible. But if you assume that molecules jiggle around at random, you can use Monte Carlo method to calculate the properties of a liquid based on the statistics of this random jiggling.

Teller suggested you could generate a random sample of configurations of molecules consistent with a given energy and temperature and average over them to get quantities like density and pressure. In physics, the relative probability of finding a configuration of molecules at a temperature τ with an energy E is given by a "Boltzmann factor" $\exp(E/\tau)$—so you could make this fast by sampling only high-probability configurations.

Marshall realized that if you take each molecule and move it about in a prescribed way—a "Markov Chain," in which each state is produced by a rule depending only on the last state—then you could prove this would generate a whole chain of snapshots of molecular arrangements, one after another like frames of a movie and statistically distributed according to their true probability. This way, you could simulate a whole range of likely configurations of molecules that would contribute significantly to their chemical behavior and estimate the liquid's properties just based on these, without worrying about extremely improbable arrangements (say, piling them all in a corner). With this, a new approach to computational physics was born, and the team were in the right place to implement it. Although the MANIAC[5] computer at Los Alamos was in great demand, its director, Nick Metropolis, let Arianna and Marshall have the midnight shift, in return for being lead author of their publication. Edward Teller's wife, Augusta "Mici" Teller, attempted a first version of the code, but it was completed by Arianna, who was by that point expert in programming the MANIAC. The Rosenbluths then worked together extensively on interpreting the scientific results. The final paper, "Equation of State Calculations by Fast Computing Machines," lists the authors as Metropolis, Rosenbluth, Rosenbluth, Teller, and Teller (sometimes called MR^2T^2), and the algorithm therefore became known as the *Metropolis algorithm*—after the only one who didn't contribute directly to its development. The many variants of the Metropolis algorithm are now simply known by their core elements, under the umbrella term Markov Chain Monte Carlo.

The essential elements of MCMC are all present in the MR^2T^2 paper. They tracked the x and y positions of 224 spherical molecules in a 2D square box.[6] They then picked one molecule, generated random numbers ξ_1 and ξ_2 between −1 and 1, and proposed moving the positions:

$$x \to x + a\xi_1$$
$$y \to y + a\xi_2.$$

Then they calculated the energy E using the formula appropriate to the forces included (such as electrostatic repulsion), and another random number ξ_3 between 0 and 1. If the move resulted in a lower total energy, it was always accepted, but higher energy states were also accepted with a small probability:

$$\xi_3 < \exp(\Delta E/\tau).$$

Because the algorithm is able to take a random walk to explore less-probable regions, it can explore the full range of states of a real liquid with the correct probabilities for each state. This algorithm remains one of the main ways that the properties of materials are calculated, a cornerstone of condensed-matter physics—but the main modern applications of MCMC were barely envisioned at Los Alamos. In the emerging field of data science, the algorithm touches everything from the Big Bang theory to social media surveillance.

In 1970 the Canadian mathematician W. K. Hastings realized the MCMC algorithm, so good at randomly sampling configurations of molecules compatible with physical parameters, could be used to sample scenarios compatible with observed data in any field of science.

Suppose you have many imprecise measurements of the position of a comet over time—you don't just want to know the orbit that best fits the data, but the full range of orbits that are consistent with it. This way, you can determine the margin of error for landing a probe—or find out the probability that it will strike the Earth. MCMC allows you to take data and explore the *posterior distribution* of models conditioned on the data. Instead of the xy

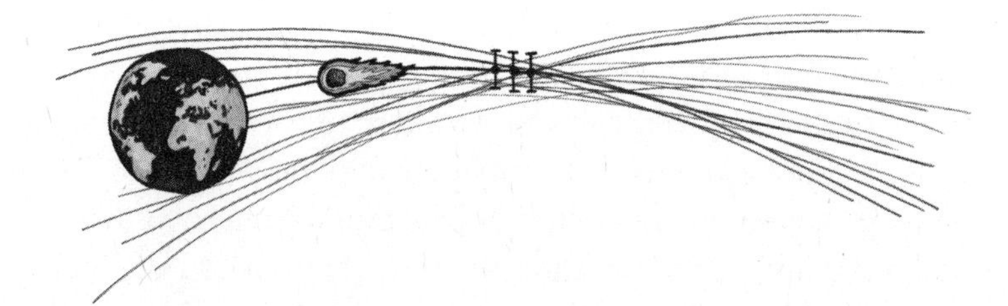

A new comet is discovered heading toward the Earth—and you only have a few uncertain data points to determine its trajectory. Using the Markov Chain Monte Carlo algorithm, you can fit many models to these data, finding that some hit the Earth, and some don't. MCMC quantifies uncertainty in models like this across every field of science.

positions of molecules, you instead take a set of model parameters (the orbital eccentricity and orientation of the comet) and do a random walk, just like before. You accept a proposed step if it makes the model more consistent with the data, or you accept the step with nonzero probability if the model is less consistent. By focusing on probable scenarios but also including less likely ones, you can explore the range of possibilities.

With Moore's Law offering ever faster computers, it has been possible to use MCMC for increasingly complicated models and volumes of data. Since 1990, hundreds of versions of MCMC (such as "Stan," named after Ulam) have made it straightforward not just to fit models to data—any optimizer can do that!—but to quantify our uncertain knowledge. In my own field of astronomy, MCMC is how you answer questions such as: Are the data really precise enough to say this? Do supernovae in distant galaxies imply the existence of dark energy? Is the universe flat? With MCMC, the Nobel Prize–winning discoverers of gravitational waves could carefully measure the masses of colliding black holes. Insurers use MCMC to assess financial risk, and Nate Silver used it to predict from murky polling data that while Hillary Clinton was favored to win the 2016 election, Donald Trump still had a good chance.

The last of the MR^2T^2 team, Arianna Rosenbluth, died of COVID-19 complications in Los Angeles just after Christmas

2020, even as epidemiologists around the world applied MCMC to infer COVID infection rates from incomplete data. Like one of her simulations, her career traces a representative sample of the triumphs and tragedies of her discipline. As part of the very first generation of computer programmers, she was also one of the first women pushed out of the field. When she was working (painstakingly, in assembly language), coding was seen as women's work, but over time the field came to be increasingly dominated by men. Despite her auspicious start, Arianna retreated from physics to care for her children, while Marshall continued on to an acclaimed career studying nuclear fusion. They divorced in 1978. Arianna never wrote another paper after those days at Los Alamos. In her later years she declined to be interviewed about her career, and her contribution was not widely known even to experts. Computer science is only now beginning to recognize the achievements of the early women pioneers, and Arianna Rosenbluth must be counted among the most important.

3

Jean Sammet and the Code That Runs the World

Claire L. Evans

Jean Sammet wore horn-rimmed glasses, stood ramrod straight, and looked the world dead in the eye.[1] Born in 1928, Jean was a math whiz. The Bronx High School of Science didn't take girls, but she excelled at her public school and eventually at the University of Illinois, where she earned a master's degree in mathematics in 1949. But when she graduated and scanned the classified ads for a job, she was quickly disheartened. In the 1950s, job listings were still separated by gender. On the women's side were openings for clerks, teachers, and housekeepers. Jean flipped the page and scanned the job listings for men.

Her strategy worked. By 1955, Jean was working as a mathematician in upstate New York, at Sperry Gyroscope, a company that produced navigation equipment and radar systems for the US Navy. One day, Jean's manager asked her, "By the way, do you know we're building a digital computer?" She was familiar with analog computers—she'd even trained on a punch-card accounting machine once, as a trainee actuary—but she spent her days at Sperry working on mathematical analysis for submarines and didn't quite know what digital computing meant. The manager said, well, digital computing was the wave of the future. "Do you want to be our programmer?" he asked.

She blinked. "What's a programmer?"

"I don't know," he responded. "But I know we need one."[2]

That was enough for Jean. As a woman, she figured she'd never be allowed on a submarine. But a digital computer was another thing entirely.[3]

"Programmer" was a job so new the classified ads never listed it at all. There were no books on the subject, no tutorials, and no instruction manuals. The new computing machines were the first of their kind. More often than not, their programming had come as an afterthought; the engineers on the Sperry project, Jean remembered, "somehow thought that machine was going to run itself."[4]

Jean bootlegged herself an engineering education as she learned to toggle binary programs into the computer by hand. Even the basic loader program took three days to enter into the machine. But programming, as it turned out, was like doing jigsaw puzzles. Jean delighted in seeing all the pieces come together and *work*.[5] She became the head of a growing department, running a shop of new hires, and when Sperry Gyroscope merged with one of its competitors, Remington Rand, she met Remington Rand's senior programmer, Grace Hopper. Hopper—who liked to call herself "the third programmer of the first computer"—had made a similar blind leap from mathematics to programming after being enlisted, by the Navy, to run the Mark I computer at Harvard during the Second World War. Like Jean, she'd had to teach herself the job.

The two became close. Though they were decades apart in age, they faced similar problems at work—both technical and social. Whenever she could, Jean would take the afternoon train to Philadelphia from her office in Great Neck, New York, to run programs on Sperry Rand's UNIVAC I computer, serving as a beta-tester in Hopper's programming division.[6] Jean had written some basic code for the Sperry computer, but Hopper introduced her to the idea of high-level programming languages. Only a few years earlier, Hopper had created a new intermediary between human and machine: a *compiler*, a string of code capable of compiling previously written computer subroutines into more sophisticated programs. It was one of the first programs capable of writing itself, a quantum leap in the field—and a lifesaving convenience for overworked programmers.

Before software was an off-the-shelf product, programmers like Hopper and Sammet were responsible for creating, debugging, and maintaining custom software installations for each client. Most everything was done in tedious machine code. And once completed, programs were essentially inscrutable to anyone but the "shadowy priesthood"[7] who had created them. At conferences, Hopper had been pushing the idea of "automatic programming": the notion that programmers should be able to step above the machine level to code with a higher level of abstraction. Proponents of automatic programming were known in their industry, somewhat derisively, as "space cadets."

After the Second World War, the electronic computers that had been developed in secret to run ballistics calculations were put to work in calculation-intensive industries like insurance and aviation, as well as in the payroll departments of large corporations and government agencies. As a result, the computing industry exploded. New machines proliferated, designed by companies like IBM, Sperry Rand, Sylvania Electric Products, and Honeywell. Not only were there not enough qualified programmers to make software for all the new computers, there were too many computers to make software *for*. This period is known by historians as the "software crisis." It was interpreted as an industry-wide lack of programming talent, but it was also a crisis of standards. Computing needed a shared vision, and a shared language, for the future. Jean Sammet would play a major role in shaping that language.

In the late 1950s, a programmer at the Burroughs Corporation named Mary K. Hawes saw the need for a single shared programming language that could be used on any machine—an *interoperable* language, created with the lay user in mind. The first meeting on the subject was held at the University of Pennsylvania Computing Center in April 1959; the small group of experts in attendance, including Hopper, outlined their goals for an industry-wide effort and put together a list of participants they wanted to enlist in the project. The second meeting, hosted by the Department of Defense—which operated 225 computing installations across the country and was beginning to develop a programming language of its own, AIMACO—was held the following May, at the Pentagon, and was attended by representatives from seven government organizations and 10 computing manufacturers, including Burroughs, GE, Honeywell, IBM, Sperry Rand, and Sylvania Electric Products—by that time, Jean's employer. The Department of Defense gave the group a word-salad acronym worthy of any military operation: CODASYL, or the Conference on Data Systems Language.

In order to keep up with their rapidly growing industry, CODASYL needed to move quickly. Several companies were already working on their own programming languages; these efforts needed to be nipped in the bud. At the same time, the task was monumental, potentially requiring years of development. The group settled on a tiered approach, delineating three committees. The Short-Range committee would examine existing languages and suggest an initial "short-range composite approach" drawing from the best of those languages. An Intermediate-Range committee would build on those findings, leaving the Long-Range committee with plenty of time to develop a finished product.

This plan went sideways almost immediately. As the Short-Range committee began work on its initial specifications, it became apparent that any interim solution they created would be permanent. In those days, it was both costly and time-consuming to implement a new programming language: Grace Hopper ballparked nearly 50 "man-years" and $945,000—worth close to $9 million in 2021—for Remington Rand alone to make the switch.[8]

In order to meet the six-month deadline for their official findings, the Short-Range group decided to appoint a smaller subcommittee from within their ranks to finalize the work. Jean Sammet volunteered immediately. ("She was not one to sit on the sidelines," remembers Dr. Tim Bergin, a friend and longtime colleague of Sammet's, and "if there was no leader in the room, she'd stand up."[9]). The final six-person subcommittee—which also included Gertrude Tierney from IBM—spent two weeks holed up in a New York City hotel, pulling all-nighters, before finishing their specifications for an interim language in November of 1959. It became known as the Common Business-Oriented Language, or COBOL.[10] Betty Holberton, a former ENIAC programmer, checked the new language line-by-line before the Department of Defense printed the final specifications for COBOL that following January.

Thanks to some clever politicking by Hopper, the Navy adopted COBOL as a standard, forcing its manufacturing contractors to build machines that could run the language. And soon the entire computing industry reoriented itself toward the new language, which became so entrenched in computing infrastructure that the Long-Range Committee was never even formed. By 2000, 80 percent of all code *on the planet* was written in COBOL. Even today, almost half of American banking systems run COBOL, and 95 percent of all ATM transactions rely on it to function.

COBOL is not a beloved programming language. The Dutch computer scientist Edsger W. Dijkstra famously called teaching COBOL "a criminal offense."[11] These days, scarcely anyone teaches it, which means there is a global shortage, yet again, of programmers. Although sensitive infrastructure worldwide relies on COBOL to function, its ongoing maintenance is continually deferred due to lack of resources and political appeal. During the early days of the COVID-19 pandemic, state unemployment systems built on COBOL were strained to the brink, with few qualified people to fix them. As the technology historian Jeffrey R. Yost points out, in our technoculture, "*Innovation* is revered, and *maintenance* is not."[12] When we ignore the vital importance of maintenance, we are blind to what actually holds our world together. We are blind, also, to the maintainers.

Jean became the first woman president of the Association for Computing Machinery (ACM). She wrote what is widely considered to be the textbook on the history and fundamentals of programming languages. In 1966, seven years after the first CODASYL gathering, she presented a paper to the ACM in which she proposed that programming languages would someday disappear, allowing programmers—and lay users—to communicate with computers using words. At the time, demand for programming talent far outpaced supply. Sammet prescribed two remedies: "to make the professional programmers themselves more productive by giving them better tools," like compilers, or to eliminate professionals entirely, making "the computer available to everyone who has a problem he wishes to solve."[13]

As utopian as this may have seemed in 1966, in the decades that followed, Jean's predictions came into focus. Today, we do have better tools; we speak with machines without a second thought. Much of this is due to a foundation built by Jean Sammet and her peers, which proved hardier than she could ever have imagined. When she was 72 years old, Jean watched programmers around the world band together to shepherd her work into the new millennium. COBOL survived the Y2K crisis; it survived Jean. It survives still.

4

Spacewar

Collaborative Coding and the Rise of Gaming Culture

Arthur Daemmrich

During the winter of 1961–1962, a loosely connected group of young MIT employees, students, and associates coded the space battle simulator Spacewar and in effect launched gaming culture.[1] Spacewar is easy to learn but challenging to master, as I have discovered while playing it on a variety of platforms. In it, two players control spaceships—one the needle and the other the wedge—and engage in a dogfight. In addition to trying to shoot each other, players face several challenges: fuel and photon torpedoes are limited, multiple hyperspace jumps increase the probability of their ship exploding, and misjudging a gravity well in the center leads to quick immolation.

Spacewar came about through the fortuitous alignment of eight young computer-savvy individuals, a leading-edge mainframe, and an environment supportive of risk-taking and collaboration. It started with an idea formulated in the summer of 1961, when Steven Russell, Wayne Wiitanen, and Martin Graetz rented an apartment together in Cambridge, Massachusetts. They jokingly named themselves the "Hingham Institute Space Warfare Study Group" and envisioned taking part in space battles similar to those in their favorite science fiction books and low-budget films.

The opportunity to go from abstract concept to an interactive space game came in September 1961, when Digital Equipment Corporation (DEC) gave a PDP-1 to MIT's Electrical Engineering department. DEC referred to it as a "programmed data processor" to differentiate from custom built "computers" that filled entire rooms and cost $1 million or more.[2] The PDP, by contrast, sold for $120,000 and fit in a footprint of 17 ft^2. MIT's PDP-1 came under the supervision of Jack Dennis, then an assistant professor in Electrical Engineering. It was installed in Building 26, which famously housed an IBM 7090 (accessible only to trained operators) and the TX-0, an early fully transistorized computer built by the military-funded MIT Lincoln Laboratory. When a cathode-ray tube display was added and Dennis offered members of the Tech Model Railroad Club (TMRC) access to the PDP-1 in exchange for their creating an assembler, the stage was set.

Coding Spacewar proceeded in fits and starts. Russell, who had worked on LISP projects for the mathematician John McCarthy (cofounder with Marvin Minsky of the MIT Artificial Intelligence Lab), agreed to code the much-discussed "demonstration" of the PDP-1's capabilities. But as his nickname "Slug" might suggest, he delayed starting, citing a lack of key mathematical tools. Alan Kotok secured them from DEC and presented them to Russell with a flourish at a gathering of the TMRC members: "Here are the sine and cosine routines. Now what's your excuse?"[3]

Russell then worked out the game's core features, notably the two spaceships, with thrust and firing from switches on the PDP-1's front panel. Ships leaving one side of the screen returned on the other. He also added a bright star in the center and random other stars in the background.

MIT undergrad Dan Edwards suggested the star should exert gravity.[4] The gravity function, however, took a toll on the program's execution speed. So Edwards pioneered a "run-time compiler trick" that loaded the spaceship outlines when the program was initialized, speeding up the main program loop.[5] Both ships accelerated toward the star in a realistic manner, making the game more challenging and setting the stage for strategies such as a slingshot maneuver to come up behind an opponent.

Steven Piner, an MIT student and member of TMRC, wrote a text editing program that made revisions to Spacewar and subsequent programming of the PDP-1 considerably easier. Jokingly named Expensive Typewriter since it mimicked the function of a $395 IBM Selectric typewriter on a $120,000 computer, Piner's program has since been celebrated as the first word processing program written for a general-purpose computer.[6]

Robert Saunders saw players jostling for space in an area meant for a single operator, became concerned about damage to the PDP-1, and built custom controllers. Made of wood with a Bakelite top, each had a firing button and two small levers, one for right-left rotation and the other for acceleration when pulled back and hyperspace when pushed forward.[7]

Peter Samson, an MIT undergraduate and member of TMRC, wrote a program for the background stars, as well as assembler code to generate the plot. Samson's star map displayed some 1,000 stars including major constellations. It also gradually shifted over time to mimic the view from earth. Samson's contribution was nicknamed the Expensive Planetarium.

Martin Graetz, who had been enrolled at MIT and then held various programming jobs at Harvard and MIT, coded hyperspace as an escape option. When initiated, the spaceship disappeared with a "Minskytron signature," a curved series of dots akin to a Bohr atom.[8] Considering it a bit of a cheat, Graetz intentionally made it unreliable with a limit on jumps and the risk of reappearing in the sun.

By March 1962, Spacewar was for all intents and purposes finished. Both word and copies of it spread quickly across the academic computing community as programmers carried or mailed the program on PDP-1 paper tape to other computer labs. Russell famously brought the game with him when he moved to California in mid-1962 to work at the Stanford Artificial Intelligence Laboratory.

The game also was distributed by DEC with PDP-1s sold after 1962. Engineers used it to test PDPs prior to shipping and left it in core memory to demonstrate that the computer worked once installed. DEC even published a small brochure in 1963 that described Spacewar as a demonstration of the PDP's capabilities, anticipating befuddled responses to seeing a game on a device purchased by corporations or universities for process control, data analysis, and other nongaming purposes.[9]

As a game, Spacewar was fun to play. As a test of the PDP-1, it demonstrated the computer's processing speed, graphics capabilities, and memory. As an innovation, it symbolized a shift from computing being in the hands of priest-like technicians operating massive computers to enthusiasts programming and hacking, sometimes for the sheer joy of it.

In 1971, a public version of Spacewar was set up by recent Stanford alum Bill Pitts and his friend Hugh Tuck when they purchased a PDP-11 for $12,000 and installed it into two arcade cabinets. Each had two mounted joysticks, and play cost ten cents (with three games for a quarter). The game was nearly identical to Spacewar, though it was called Galaxy Game. It remained in the Stanford Tresidder Memorial Union for eight years.[10]

Several months later, Nolan Bushnell and Ted Dabney, cofounders of Syzygy Engineering (renamed Atari in 1972), developed and installed the arcade game Computer Space.[11] In it, a single player operated a rocket that battled a pre-programmed pattern of two saucers with a fixed star map background. Though often described as derivative of Spacewar, it had clear differences. In Computer Space, the player competes against the computer, the game runs with a countdown timer, and there is no sun with a gravity well in the center of the screen. The game worked, but Computer Space was expensive to build, and play-

```
dap hxt
isp .
jmp hp2-1
dzm hp1 i
jmp hxt
lio mx1 i
dio .
lio my1 i
dio .
lac ml1 i
dap .
random ran
add mx1 i
dac mx1 i
add my1 i
dac my1 i
```

A selection of code from Spacewar's Minskytron "Hyperspace signature," in which a player sends their ship into hyperspace, often as a last resort.

ers found gameplay difficult due to its programming and the configuration of the controls. It was displaced by other games during the 1970s, starting with Pong in 1972. By the time Space Invaders took off in 1979, the arcade run for games directly inspired by Spacewar had passed.

In 1978, Atari released a Space War game for its increasingly popular Video Computer System (renamed the Atari 2600 in 1982) home gaming platform. In the clunky Atari version, players could choose among some 17 game variants; the first seven were low-quality clones of the original Spacewar, while the next 10 were single or two-player games involving docking with a "space module."

While neither Computer Space nor the Atari version from 1978 were especially popular, Spacewar did have some lingering influence on successful video games of the late 1970s and early 1980s, notably Asteroids, Defender, and Missile Command. A collapse of the video game industry in 1983 followed by a new era of games featuring characters and more complex gameplay seemed to signal an end to Spacewar's run.

But then came emulation. Starting in the mid-1990s, a combination of increased desktop computing power, faster Internet data transmission, and popularity of "retro" gaming gave rise to emulators.[12] Spacewar attracted several detailed emulations, notably Norbert Landsteiner's "Masswerk" project. Among other details, Landsteiner recovered source code for the hyperspace Minskytron signature and analyzed of the original Spacewar code in detail.[13]

In 2005, a team of volunteers supported by the Computer History Museum restored a PDP-1 and put it on display so that museum visitors could play Spacewar.[14] Russell and Samson were themselves part of the attraction, volunteering regularly throughout the 2010s and early 2020s. Russell observed, "there are no outstanding user reports of crashes. There are no user complaints outstanding. It's fifty years old, it's still running, and support is still available."[15]

Historically, the Spacewar code itself was never truly final. While Russell identified a "finished" version in 1962, it was modified with each port to a new system. Rewrites in the 1960s with the release of each DEC PDP model involved hours of coding, and programmers often took liberties to modify the game. Versions from the 1970s and 1980s suffered from poor graphics and uninspired gameplay, but also included some creative new features. By contrast, the contemporary emulated versions have focused on the graphics, backdrop, and gameplay of the 1962 version. With no way to go back in time to MIT with a freshly installed PDP-1 and people standing around coding and playing Spacewar when they are supposed to be doing computing's "real work," the emulations are technical, not social in nature. But it was the social process of coding that created a uniquely innovative environment for computing in the 1960s and 1970s. Not an environment without flaws, it must be said—most notably around issues of gender and race since women and people of color were absent or marginalized.[16]

Spacewar had several unique features that contributed to its success but were irreproducible in subsequent programming history. First, it was coded through a voluntary and largely asynchronous process. The full group was never together in one location at the same time until a 2018 retrospective at the Smithsonian Institution.[17] Second, for about a decade, the first time most people saw Spacewar was also the first time in their lives that they saw an interactive graphic computer game. Although it was not the first computer or video game, previous demonstration projects like Higinbotham's oscilloscope-based Tennis for Two were unknown to most audiences. Third, the coders had no commercial aspirations associated with the game; it was public domain and predated intellectual property disputes regarding copyrights and patents on software.

Yet, like a surprising number of other programs, the Spacewar code lives on far beyond its inventors' expectations and even the expected operational lifespan of its host machines. For programs such as COBOL-based unemployment software, long-term use demonstrates the robustness of code but also failures to invest in new systems. For a program like Spacewar, long-term use demonstrates the innately human joy of play and the desire of several generations of coders to preserve and celebrate their own history.

5

BASIC and the Illusion of Coding Empowerment

Joy Lisi Rankin

During the first half of 1964, two college-age White men, John McGeachie and Michael Busch, devoted hours to computer programming. So much time, in fact, that McGeachie was known as 225, short for the GE-225 mainframe computer for which he was responsible, and Busch was known as 30, short for the GE Datanet-30 computer that he programmed. They were students at Dartmouth, an elite, overwhelmingly White, Ivy League college that admitted only men as undergraduates, and they were coding a new computing network. In the early 1960s, McGeachie's and Busch's access to technology was extraordinary.

In the 1960s, most mainframe computers ran on batch processing. Programs were communicated to the machine through inputs known as keypunch cards. Holes punched in the cards communicated numbers, letters, and symbols to the computer. One program often consisted of many cards. At the time, managers sought to keep computers running as much as possible—they were quite expensive, and organizations wanted to get their money's worth—so individual programs were grouped together and run in large groups, known as batches. For example, before Dartmouth acquired its own computer, Dartmouth professor Tom Kurtz made daytrips by train to use the MIT computer, carrying with him a box full of punched cards encoding his and his colleagues' programs: economics models, physics simulations, mathematical equations.

Typically, a computer operator handled the batch input process, as well as retrieving output such as printouts. As a result, someone who wanted to create and run a computer program had no interaction with the computer system itself—and they could wait hours or days for the results of running their program. This meant that the several thousand computers in the United States in the early 1960s were out of reach of nearly everyone, especially young people. Even the computers installed at universities were the province of a handful of faculty and graduate students. That would soon change.

The men at Dartmouth sought to challenge those limits of accessibility and batch processing. Math professor John Kemeny persuaded the trustees of the college that computing would be essential for Dartmouth students as the future leaders of American science and industry. His fellow math professor Kurtz envisioned a system where all students would be able to access computers directly, without the delays and middlemen of batch processing. Kurtz also imagined that computing would be freely available to students as part of their college experience like unfettered library access—being able to browse and pull books directly off the shelves, rather than submit a ticket for someone else to retrieve a book. Finally, Kurtz believed that Dartmouth could accomplish this by building a time-sharing network.

Time-sharing was a new form of computing in the 1960s. Time-sharing *sounds* like computer users were signing up for blocks of computing time: Alice gets 15 minutes, then Bob gets 15 minutes after Alice. But it actually means programming a mainframe computer to share its own time and computing resources among multiple programs running at the same time. In effect, this meant that multiple people could sit at individual terminals connected to one mainframe and write, run, and debug their programs at the same time.

On the Dartmouth network, the individual terminals were teletypewriter terminals that had been developed for telegraphy. They looked like old-fashioned typewriters with large printers built in. A user saw their program as they typed on the teletype, and the computer communicated results to them by printing on the teletype. Telephone wires connected teletypes to the mainframe. This meant that terminals could be—and were—located far from the network's mainframe, even in another state or halfway across the country.

In May 1964, the Dartmouth College Time-Sharing System, the early personal and social computing network that McGeachie and Busch helped program, was launched with the simultaneous and successful run of two BASIC programs. BASIC was Beginner's All-purpose Symbolic Instruction Code, a computing language developed at Dartmouth under the guiding principle that it should be easy to learn and use.

We don't know exactly what those lines of BASIC code were. We don't even know who ran the two programs.[1] But we know now that for three reasons, those BASIC programs made America's digital culture possible by spreading personal computing far, fast, and wide. The first and second reasons are fairly well known: the revolutionary accessibility of Dartmouth's computer network and the radical ease of BASIC. The third reason is the most important, yet has been overlooked: how BASIC limited paths and possibilities.

Although building a computer network for undergraduate use was visionary in the 1960s, it would not have been nearly as successful if not for BASIC. BASIC and Dartmouth's network—and the rapid uptake of both—were inseparable. Computing languages prior to BASIC, such as COBOL and FORTRAN, had been developed for scientific, research, and business purposes. They were not known for being easy to learn or user-friendly. FORTRAN's name came from FORmula TRANslation, reflecting its intended use for math and science computing.

In 1967, a student at Williams College created a program to score ski jump competitions—a challenging task that took a team of faculty and students over three hours by hand. The Williams student wrote his program in FORTRAN to run on an IBM. He spent 50 hours writing it. Meanwhile that same year, an instructor at Vermont Academy created a program to score an entire ski *meet*—ski jump plus cross-country, downhill, and slalom. The Vermont instructor wrote his program in BASIC to run on Dartmouth's network. He spent 10 hours writing it.

Compared with languages like FORTRAN or COBOL, BASIC was much faster and easier to learn. BASIC's commands—including `IF-THEN`, `LET`, `PRINT`, and `READ`—more closely resembled everyday English. At Dartmouth, the combination of BASIC and the time-sharing network enabled students to quickly write and debug short programs, to experiment, to not be afraid of making mistakes, especially because they could see the results of their programs in seconds or minutes, not days or weeks. They used BASIC for their coursework and to write letters home. They produced computer art, simulated slot machines, and programmed and played games including chess, checkers, poker, and slalom skiing. By 1968, 80 percent of Dartmouth students regularly used the network and BASIC.

In that way, BASIC offered the illusion of coding empowerment. Consider the opening of this essay: sometime in May 1964, two men sat in front of two teletypes at Dartmouth, and they successfully ran simultaneous BASIC programs on the college's brand-new time-sharing network. The fact that they were young White men at an elite, predominantly White college, is central to this story, not incidental.

During the 1960s, many women and Black people worked in computing. Before World War II, a computer was a person who performed mathematical calculations. Computers worked in business and scientific settings, and when computers became machines, many women worked *with* computers: writing programs, translating business needs to computer applications as systems analysts, operating keypunches and mainframes, and filling similar roles across industries and academic disciplines.

A 1967 issue of *Cosmopolitan* magazine with the headline "The Computer Girls" celebrated computing as "woman's work." In

Hidden Figures, the journalist Margot Lee Shetterly documents how she "can put names to almost 50 black women who worked as computers, mathematicians, engineers, or scientists at the Langley Memorial Aeronautical Laboratory from 1943 through 1980."[2] Likewise, the archivist Arvid Nelsen identifies at least 57 Black Americans working in computing between 1959 and 1996—just from the "Speaking of People" column in *Ebony* magazine.[3] As Claire Evans documents in her essay in this book, well-known women like Jean Sammet and Grace Hopper were not exceptions in early computing. Rather, they embodied the fact that early machine computing was a feminine field.

That shifted during the last decades of the twentieth century, when computing gained prestige in the United States and the United Kingdom by becoming the realm of affluent White men.[4] When Kemeny sold Dartmouth trustees on the idea that computing was essential knowledge for the future American leaders whom Dartmouth was producing, he was associating the power of computing with both the Whiteness and the maleness of the college. Requiring all first-year students taking math courses to successfully write a BASIC program further cemented the relationship among computing, Whiteness, affluence, and power at Dartmouth.

When other schools and universities around New England expressed interest in connecting to Dartmouth's network during the 1960s, Kemeny and Kurtz happily acquiesced. In fact, the college even secured a National Science Foundation (NSF) grant to support connecting 18 high schools around New England to the Dartmouth network. Some high-schoolers regularly woke at four in the morning to use the network.

But access to the Dartmouth network was by no means equal, and it was generally young, wealthy, White men who benefitted the most. Among the high schools connected to the Dartmouth network as part of the NSF Secondary Schools Project, the coed public schools—all predominantly White—had only 40 hours of network time each week. By contrast, the private schools—which were all male, wealthy, and almost exclusively White—had 72 hours of network time each week. In these years before the expansion of educational opportunities for American women, high school boys were still enrolling in many more math and science classes than high school girls. And it was in those math and science classes that they gained access to computing. During this decade of the Civil Rights Movement, Americans were reckoning with the myriad ways in which their public schools were separate but by no means equal. BASIC traveled in an American educational system that was already segregated by gender and race, so it ultimately amplified inequity in terms of computing access.

Kemeny and Kurtz decided to make BASIC's source code freely available so that BASIC could be (and was) implemented across many different makes and models of computers and networks. BASIC programs were stored on networks, shared in handwriting or by word of mouth, and soon circulated in books and informal newsletters, including the popular *People's Computer Company*. BASIC originated the idea that programming was something that just about anyone could do. And the echoes of that unexamined assumption perpetuate the pernicious myth today that all you need to do to succeed in tech is learn how to code.[5] BASIC made learning to code easy—but for whom?

6

The First Email

The Code That Connected Us Online

Margaret O'Mara

"A new command should be written to allow a user to send a private message to another user which may be delivered at the receiver's convenience." This suggestion appeared in a "programming note" written by managers of MIT's computer time-sharing system sometime in late 1964 or early 1965, a document so informal that no one bothered to affix the precise date at the top. Neither the authors, nor the young programmers who wrote the `MAIL` command in response six months later, realized the communication revolution they were helping to start.[1]

The first all-digital computers were designed to be machines of calculation, not correspondence. These mainframes were giant in every respect: of massive proportions, hugely expensive, accessible only to the largest government agencies and corporations. As computer makers chased the commercial market during the 1950s, they optimized for the very large batch data-processing operations demanded by corporate customers, designing for one user, and function, at a time.

The scale and singularity were incompatible with the needs of academic researchers, who worked collaboratively and conducted more incremental, intermittent work of program development. In 1955 MIT computer scientist John McCarthy began conceptualizing a system where multiple users could share computer time. The product of those early musings, MIT's Compatible Time-Sharing System (CTSS), launched in 1961.[2]

As in real life, when multiple people collaborated in a computing environment, they wanted and needed to communicate with one another. Thus, as users took turns on CTSS—working at different times and often in different places—they began to leave messages on the system for their colleagues to read as they logged on later. These electronic notes were professional, not personal; they communicated system updates, or relayed questions or critiques about particular files. Yet there was no way to direct a message to one user, nor keep messages on file for later reading.

But the men and women of CTSS were too consumed with other, higher programming tasks to get around to writing the code that would add these improvements to the ad hoc messaging system. It wasn't until the summer of 1965 that two recent MIT graduates, Tom Van Vleck and Noel Morris—newly hired as university staffers and so entry-level that they shared one windowless office—decided to take on the task.[3]

The `MAIL` command was the result. It was quite straightforward, really. First, type `MAIL`; then `NAME 1` and `NAME 2`, together representing the name of the file, following CTSS's two-name convention; then `PROG 1`, the programmer who was the recipient. There was also a `LIST` option to send a message to a mailing list. (Yes, the listserv is as old as Email itself.)[4]

Mail programs quickly became common features of time-sharing systems throughout the 1960s, but they remained limited to the users of one system. The next technical challenge was to find a way for messages to travel across different computers and hosts. Enter the ARPANET, launched in 1969, funded by the Pentagon and theorized and designed by some of the same computer scientists behind academic time-sharing systems.

At first, what would become Email did not appear to be very important to the ARPANET. As J.C.R. Licklider, a chief visionary of the system, later put it in a paper coauthored with his colleague Albert Vezza, "electronic message service was a sleeper." Yet it quickly became clear that allowing users to communicate

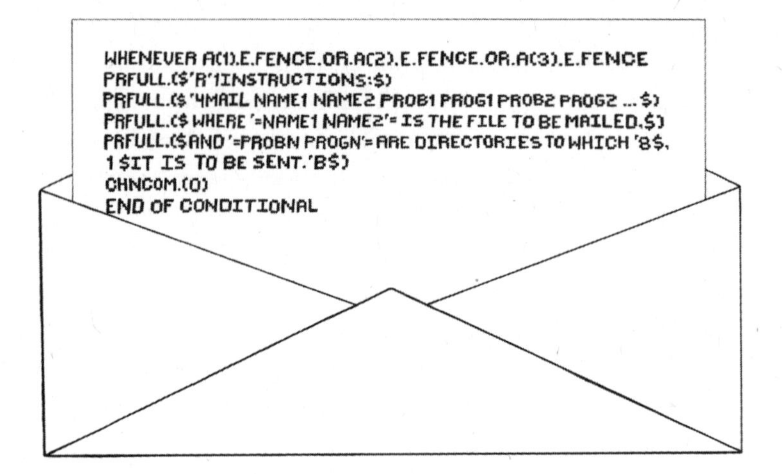

The MAIL command, which produced the world's first Email.

across the network was going to be one of ARPANET's most useful functions.[5]

This was thanks to programmer Ray Tomlinson, who in 1971 established a new protocol for ARPANET users connected via different hosts, fatefully choosing an "@" to connect a username and host name as a way to get messages across the network to its intended recipient. It was the missing link Email needed in order to scale, allowing users of different systems, in different places, to communicate as easily as if they were in the same laboratory. Even then, Tomlinson didn't take his invention that seriously; "the test messages were entirely forgettable and I have, therefore, forgotten them," he later admitted.[6]

In the total-immersion hacker culture of the early online world, Email was quick, informal, tolerant of typos, and immensely preferable to the time-wasting small talk of phone calls. As Licklider and Vezza noted, "it soon became obvious that the ARPANET was becoming a human-communication medium."

The intensely collaborative, transparent spirit of American academic computer science, and the ARPANET in particular, proved particularly helpful in rapidly advancing the state of the art during the 1970s. Email began to take on its modern form as programmers contributed suggestions and program notes in a series of openly accessible RFCs—requests for comment—posted among network users. Various RFCs suggested now-familiar Email characteristics: a "`to`" line, a "`from`" line, and the ability to forward messages to another user.

As the ARPANET grew, it began to introduce electronic mail to a wider audience. In 1976, Queen Elizabeth II christened the ARPANET's first node in Great Britain by sending an Email on the network. (Her username: `HME2`.)[7] In 1978 an overeager computer-company marketer won the dubious honor of sending the first piece of commercial spam, blanketing ARPANET users' Email inboxes with a note promoting a new model. By then, over three-fourths of the network's traffic consisted of electronic mail.[8]

In the 1980s, Email moved beyond the ARPANET's walled garden of government-funded computer science. The proliferation of desktop computers in homes and offices vastly enlarged the market of computer users, and new applications made Email accessible to this wider public. The popular online news and information service CompuServe began offering its home-computer customers electronic mail accounts in 1979. One alumnus of the University of Illinois computing center, Ray Ozzie of Lotus Software, developed the wildly popular business communication software Lotus Notes in 1986. Another University of Illinois researcher, Steve Dorner, developed the Eudora Email client in 1988 (naming it after Eudora Welty, the celebrated Southern novelist and author of the short story "Why I Live at the P.O.").

After the US government allowed the ARPANET to become commercialized in the early 1990s as the Internet, Email often became the first program a new user experienced. It was an accessible gateway into the daunting and exciting new online world.[9]

By the time the dot-com boom crested at the end of the 1990s, Email had become ubiquitous. Steve Dorner's Eudora program had 18 million users; IBM bought Lotus for $3.5 billion thanks to the galloping success of Notes. The animatronic bark of AOL's "you've got mail!" notification was everywhere from television commercials to Hollywood rom-coms. It became hard to imagine how the world ever managed without it.[10]

In the 20 years since, the Internet has grown exponentially, and Email has grown with it, although other communication media have supplanted it to some extent. Email's informal cadences have given way to even more informal text strings and ephemeral social media posts. Yet the new forms still retain the characteristics—brevity, personalization, asynchronicity—that made Email so popular in the first place.

Six decades after that first programming note posted to the internal servers at MIT, billions of Emails shoot daily around the globe, vital in their convenience, overwhelming in their abundance, making the crowded inbox an inescapable feature of modern life. Rarely celebrated and often lamented, Email fundamentally changed the way people used computers and interacted with one another. `MAIL` turned out to be not just another command; it became the code that gave the computer its humanity.

7

The Police Beat Algorithm

The Code That Launched Computational Policing and Modern Racial Profiling

Charlton McIlwain

In the early 1960s, the Black civil rights revolution raged in the streets across the United States. This quest to build a more racially just and equitable society happened right alongside the computer revolution. Soon the two fused with the advent of the Police Beat Algorithm (PBA), a software system to help police departments collect crime data and determine where to focus crime-fighting efforts—and one that that would end up deeply affecting our society from the 1960s up through the present. Why did the Police Beat Algorithm come to exist? What problems prompted the need for its formulation? Who developed it, and to what ends? The answers to each of these questions collectively tell a story about how a little-known computational experiment laid the cornerstone for what would become today's surveillance infrastructure—one that has deeply and negatively affected communities of color across the globe.

In the early 1960s, IBM topped the list of the world's leading computing companies. It innovated not only new computer hardware and systems but new ways of thinking about the computer's role and utility in everyday society. In its 1965 Annual Report, IBM president Thomas J. Watson Jr. defined the computer as essentially a problem-solving tool and aligned the company's mission accordingly.

IBM's focus on problem-solving also dictated its marketing strategy. The company's marketing representatives didn't peddle prepackaged products. Rather, they engaged leaders in every major industry—from banking to transportation to the military—and simply asked, "What problem do you have?" Then, they promised to marshal IBM's research and development strength to build customized solutions for its customers—solutions that could be broadly applied and widely scaled.

While IBM labored to market new computational solutions to social problems, uprisings materialized across the United States. In 1964 alone, so-called ghetto riots broke out in places like Harlem and Rochester in New York; Philadelphia, Pennsylvania; and Dixmoor, Illinois. These uprisings captivated the nation, as did the rampant White violence against those who marched for civil rights across the South. In a speech to Congress on March 15, 1965, President Lyndon Johnson proclaimed that America's "Negro problem" was America's problem. Citizens across the United States identified this fracture in "race relations" as the nation's most pressing dilemma.

For most White Americans, however, the urban uprisings that plagued the nation revealed Black Americans' penchant toward violence and criminality—so much so that President Johnson's White, Southern constituents thought solving America's crime problem should be his government's top priority. Heeding their agitation, Johnson, on July 23, 1965, formed the President's Commission on Law Enforcement and the Administration of Justice. The Commission's charge was to study the causes of, and find solutions to, America's crime problem.

Just 19 days later, one of the most deadly and costly uprisings erupted in Watts, Los Angeles. One too many incidents of police brutality at the hands of the Los Angeles Police Department set off six days of unrest. Hundreds of LAPD police officers flooded the streets. Fourteen thousand National Guard troops stormed the city. Law enforcement killed 34 Black residents and injured thousands more. More than $40 million worth of property was damaged during the siege.

Through the Watts uprisings, Black America sent a message to White America: We're fed up. We're tired of racism, discrimination, and police brutality. White Americans, however, saw Watts as confirmation of their prejudiced belief that Black people are lawless and violent. For the President's Crime Commission, White America's vision of the Watts uprisings put a face to the problem the president called on them to solve—a problem that they felt required an extraordinary remedy. They found great potential in the new computing technologies that had already revolutionized war and national defense. Computing held so much promise that in the spring of 1966, following the Watts uprisings, Johnson added the Science and Technology Task Force to the Commission to introduce new computational solutions to crime. The president justified the task force's work by pointing to computing technology's success in war, national defense, and space exploration:

> The scientific and technological revolution that has so radically changed most of American society during the past few decades has had surprisingly little impact upon the criminal justice system. In an age when many executives in government and industry, faced with decision making problems, ask the scientific and technical community for independent suggestions on possible alternatives and for objective analyses of possible consequences of their actions, the public officials responsible for establishing and administering the criminal law . . . have almost no communication with the scientific and technical community. More than two hundred thousand scientists and engineers are helping to solve military problems, but only a handful are helping to control the crimes that injure or frighten millions of Americans each year.[1]

While the president and the Commission held great hope for the solutions the Science and Technology Task Force would produce, they placed their hopes more specifically in the one man whom they appointed to lead it: Saul I. Gass.

Gass was a mathematician and operations research pioneer. In 1958 he wrote the first textbook on linear programming—a mathematical modeling technique that seeks to (in large part) influence human behavior by quantifying and understanding the linear relationships between variables. Gass went to work for IBM in 1960 as project manager for the company's contract to develop the real-time computational systems needed for Project Mercury, the United States' first manned space mission. By 1965, when the President appointed Gass to lead the Science and Technology Task Force, Gass was managing all of IBM's federal system projects. By heading the task force, Gass signaled his agreement with the Johnson administration that policing was the institution best equipped to solve America's crime problem—and therefore developed—the Police Beat Algorithm.

The Police Beat Algorithm was designed to address two broad planning questions.[2] First, how should police departments equitably divide the geographic and demographic parameters of a municipal area? (Gass focused on "urban" areas based on population, crime levels, and demographic factors.) Second, how should police departments effectively deploy police resources (people, weapons, vehicles, etc.) based on these geographical divisions? Interestingly, Gass frequently highlighted the need to solve these problems in order to develop "contingency riot and other emergency plans"—a growing concern directly tied back to Watts and similar uprisings.

The Police Beat Algorithm predominantly addressed four problems associated with police operations: 1) pattern recognition, identifying crime patterns within a set of crime data; 2) profiling, associating crime patterns with probable suspects; 3) dragnetting, linking probable suspects of one crime with past crimes or arrests; and 4) patrol positioning, how to best place patrols within appropriate geographical divisions of the city based on where the most crimes take place and where known criminal suspect profiles predicted who will most likely commit those crimes and where. This is where planning problems and operational problems intersected.

The Police Beat Algorithm was designed to focus on patrol positioning. Doing so relied on one primary component—the availability of crime data—and two key computational techniques, norming and weighting. Norming refers to analyzing the data to determine "normal" and aberrant ranges of criminal activity, both across a geographical area and for particular groups of criminal suspects (White people versus Black people, for example). Weighting, in this instance, was a means to rank the severity of different crimes. For example, crimes like homicide, rape, burglary, larceny, and auto theft were weighted with a score of four, signifying the most severe forms of crimes. Some of the arbitrary—or dare I say biased—nature of these weights can be seen in the lack of weighted differentiation between crimes against humanity like homicide on the one hand, and property crimes like car theft on the other. Traffic accidents received a weighted score of two, and drunkenness, a score of one. Geographical areas were weighted by the preponderance of crimes committed within their boundaries. The crime data, the statistical norms, weights, and geographical configurations of a city all figured into the Police Beat Algorithm.

In one respect, the PBA was developed to address a problem that framed Black people—primarily those who were poor and lived in urban environments—as predominantly responsible for crime and, as a result, the problem that needed to be solved. The Police Beat Algorithm was therefore predetermined to geographically locate, isolate, and target Black and brown communities for police profiling, surveillance, and patrol and tactical unit distribution and deployment. All of the resulting data from these "solutions" could be used to forecast and predict where crime was most likely to happen in the future and allow police

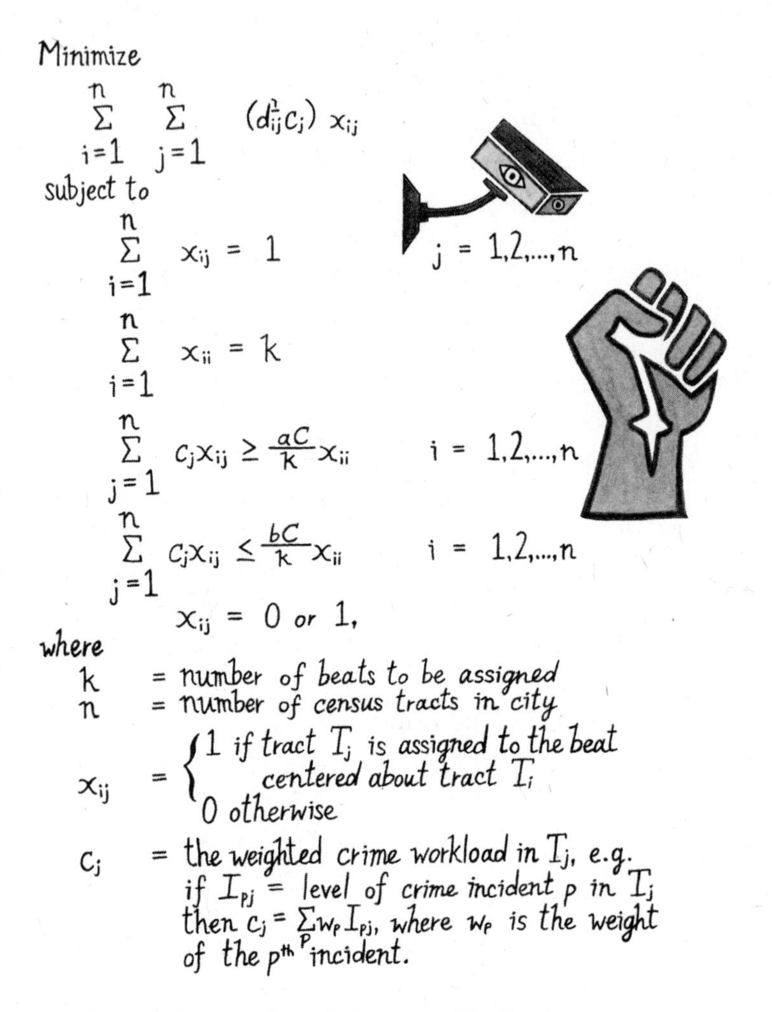

The Police Beat Algorithm, along with its computational key.

to plan accordingly. To be sure, the framing of the problem, and the configuration of the Police Beat Algorithm itself, promised outcomes that were not so much predictive of future crime as they were self-fulfilling prophesies.

Gass's PBA was essentially a proof of concept. Nevertheless, it was implemented in 1968 in the Kansas City Missouri Police Department's new Alert II Criminal Justice Information System.[3] It was through this system that the PBA's racist impact was fully realized. Kansas City's "Operation Robbery Control"[4] was just the first example of how the algorithm led police officials to make the tactical decision to concentrate police personnel and deploy weapons on what was essentially the whole of East Kansas City, which housed the vast majority of the city's Black citizens.

Ultimately, the Police Beat Algorithm became thousands of similar systems designed and built throughout the seventies, eighties, nineties and beyond. Over the decades, these algorithms have grown to include facial recognition, mobile surveillance, risk assessment, and other such tools used from local law enforcement to international security. The same logics and assumptions that motivated the creation of the PBA more than 50 years ago continue to permeate this array of contemporary law enforcement technologies. Fear of crime—still personified disproportionately by Black and brown people—continues to be greatly exaggerated, justifying exorbitant investment in developing more law enforcement technologies. Belief in the objective and infallible nature, and in the predictive power of data, continues to run rampant among technology purveyors, law enforcement personnel, public officials, and policy influencers. And stories about the disparate outcomes these technologies have on communities of color continue to roll in like a steady drumbeat. In these ways, today's law enforcement technologies are not new; they're just more sophisticated, insidious, ubiquitous, and more impactful than when the PBA was first conceived more than half a century ago.

8

"Apollo 11, Do Bailout"

Ellen R. Stofan and Nick Partridge

As Neil Armstrong and Buzz Aldrin guided the lunar module *Eagle* toward the Sea of Tranquility, Earth was watching. Apollo 11 was inching its way to a moment a decade in the making and centuries in the dreaming. At around 30,000 feet from the lunar surface, eight minutes from their planned rendezvous with history, an alarm blared. The code 1202 flashed on a tiny display—and neither the astronauts nor flight controllers knew what it meant. What President Kennedy had called the "the most hazardous and dangerous and greatest adventure on which man has ever embarked"[1] hung in the balance by a tiny thread of binary code deep in the space age machine.

It's become a techno-trope to compare the raw computing power of the Apollo Guidance Computer (AGC), the system that guided history's first (and so far only) crewed lunar voyages between 1968 and 1972, to devices we carry with us every day. Our phones, calculators, even smart watches outstrip the AGC in every measure of bit and byte.[2] Poppy Northcutt, who calculated Apollo's return-to-Earth trajectories, once told one of us wryly that the AGC was less powerful than a recordable birthday card—but as Poppy understood better than anyone, the comparison distorts the bigger picture. Comparing the AGC to an iPhone is like judging a piece of music based solely by the number of notes on the page. To understand the virtuosity of the Apollo code, it's important to understand a bit more about the unique system and circumstances for which it was written.

Each Apollo mission carried two AGC units, one in the command module (which stayed in lunar orbit and then returned to Earth) and one in the lunar module (which carried two of the three astronauts to the Moon's surface). The computers were virtually identical, and both comported to tortuous restrictions on size, weight, and power consumption. Our astronauts were guided by machines no larger than carry-on luggage in an era when most computers were the size of small houses—the AGC was roughly one cubic foot, weighed 70 pounds, and drew less power than a 60-watt light bulb. The whole system, including more than 2,000 then-novel silicon transistor chips, was sealed into an unassuming metal box tucked away in an equipment bay. In terms of software, there was no app store for the AGC—its only hardwired application worked toward one perfect moment of human-celestial contact and a safe return.[3]

The AGC's software was responsible for a wide array of functions and datasets to control the movement of two spacecraft in multiple space environments, all while tracking star positions for guidance and the motion of the Earth and Moon in their orbits. The AGC directly controlled more than a hundred systems and devices within the spacecrafts, from the sextant and guidance platform to the attitude control thrusters and main engines.

The astronauts interacted with the AGC through a display keyboard. The inputs were sorted into "verbs" and "nouns." Verbs were actions the computer could take at the astronaut's command, such as "`load`" or "`display`." Nouns were data that could be acted upon by the verb commands, such as star positions or trajectories. The setup and nomenclature were holdovers from the early development process at MIT. The top brass at NASA were concerned that the interface was clunky and out of keeping with the sleek, high-tech spirit of the program.[4] But as developers often discovered in the leadup to Apollo, astronauts are smart people, and they found the system intuitive, even elegant. Apollo 15 Commander David Scott later remarked: "It was so simple and straightforward that even pilots could learn how to use it."[5]

While a tiny fraction of the AGC's memory was random access (RAM, in modern terms) for certain tasks and procedures, the majority of the software was hardwired—literally.[6] Programs were encoded physically on rope core memory—wires running through tiny rings made of iron ferrite, which were naturally inured to transiting magnetic fields or streaking cosmic rays in cislunar space. If the copper wire went through the iron donut, the computer would read a 1; if it went around the ring, it was a 0—familiar to us today as a single binary bit.[7]

The whole works was then woven together by hand into fabric ropes, mostly by women in factories in Sudbury, Massachusetts, during the development stage, and then nearby Waltham for the missions. The weavers earned the affectionate, if diminutive, nickname "little old ladies" or LOLs.[8] (LOL, it turns out, was computer slang decades before it reemerged on the proto-Internet in the 1980s and entered the OED in 2011.) But far from dismissing their work, NASA recognized the LOLs' importance. When astronauts visited the factory to meet the women who held their lives in their hands, the weavers said, "that could be my son, so I am going to do my job as well as I can."[9]

The ingenious rope core design also meant the hardwired software could not be easily changed, corrected, or modified once encoded. Even more than the spacecraft themselves, the bespoke code for each mission had to be ready early, and it had to be flawless.

The code itself, the actual instructions for how the mission should weigh, prioritize, and combine all of the myriad variables, was written by a team from the MIT Instrumentation Laboratory (later renamed the Charles Stark Draper Laboratory, after its founder).

The programmers' notes to one another in the code read like a Talmudic commentary through the lens of 1960s slang—an instruction to reposition an antenna is labeled "CRANK THE SILLY THING AROUND" in order to safely continue "OFF TO SEE THE WIZARD."[10] Another sequence ends with "COME AGAIN SOON," and a block to control instrumentation displays was filed under "PINBALL_GAME_BUTTONS_AND_LIGHTS." In a nod to the language of the display keyboard's interface, there was even a bit of Shakespeare—"Thou hast men about thee that usually talk of a noun and a verb, and such abominable words as no Christian ear can endure to hear."[11]

Also apparent was the stress of writing thousands of lines of code on a timeline that was tight even by Apollo's standards—one section that made it into the final version was marked "TEMPORARY, I HOPE HOPE HOPE." The state of the nation and the world was also reflected—an ignition sequence is titled "BURN_BABY_BURN," an allusion to the Black Power movement, according to one young programmer. The fight for civil rights and the Vietnam War were heavy on their minds, even as they worked to deliver us to another world.

The command module's primary software was called COLOSSUS, the lunar module's LUMINARY (C for command, L for lunar). The way the two worked in concert with the hundreds of other programs, processes, and tasks was where the AGC's sophistication truly shined through—especially in the tense moments leading up to the first landing.

Although it strained every limit of early computer science, the AGC *was* limited—just like modern computers, it couldn't do everything all at once. So it had to prioritize, and it did that well. Tasks were accorded a level of importance based on the criticality of the action, the consequences for dropping or delaying it, and the power demands of the process, among other factors. Some tasks like navigation—while critical—could be interrupted or postponed by an astronaut's query, or to protect critical data. AGC always focused on the most important jobs—and the software team knew there were eventualities they couldn't plan for—and as it happened, one of the most unlikely exigencies occurred just eight minutes before the first landing attempt.

The Apollo-Saturn stack—from the rocket that sent the astronauts to the Moon to the parachutes that brought them home and everything in between—was so complex that no one person could know it all. The astronauts themselves were supported by the flight controllers, who were themselves backed up by hundreds and hundreds of engineers and systems experts.

The lunar module was made up of two stages—the descent stage, with its iconic spider-like legs and golden Kapton wrapping, which would become a lunar launch pad after landing, and the ascent stage, which held the astronauts and the rocket that would carry them back to the command module. Designed to launch from the Moon's surface during a successful mission, in an abort scenario the two would separate in space by blowing pyrotechnic bolts, sending the descent stage crashing into the Moon below and hurtling the ascent stage back up into orbit. It was a dangerous maneuver, impossible at lower altitudes, that would risk the lives of the crew and end the mission.

When the 1202 alarm interrupted Apollo 11's first and final approach, Neil Armstrong called down to Houston for help as his hand hovered over the abort handle that would snap his spacecraft in two and end history's first moonshot.

Some of the engineers wanted to abort, but in less than 30 seconds another one had diagnosed the problem. The call came from the back bench: GO! The alarm actually signaled a solution, rather than an error—the AGC code was functioning exactly as designed and had reacted so quickly the astronauts and flight controllers rushed to catch up.[12] The 1202 code signaled a condition called "Executive Overflow," caused by surplus radar data being fed into the overstretched processor. This triggered a sequence of code called "BAILOUT," which instructed the computer to drop less critical tasks and restart to save the flight data needed for landing. In this crucial moment, the computer even considered the spacecraft's display screens expendable—and they went dark. The restarts were happening so fast that the astronauts couldn't detect them. If the alarms were only intermittent, Houston determined, the mission could proceed. With their lives on the line and the whole world watching, the 1202 and related 1201 overflow alarms sounded four more times before Armstrong uttered "Houston, Tranquility Base here. The *Eagle* has landed."[13]

```
POODOO   INHINT
  CA  Q
  TS  ALMCADR

  TC  BANKCALL
  CADR  VACSSTOR   # STORE ERASABLES FOR DEBUGGING PURPOSES.

  INDEX  ALMCADR
  CAF  0
ABORT2   TC   BORTENT

OCT77770  OCT  77770       # DONT MOVE
  CA  V37FLBIT  # IS AVERAGE G ON
  MASK  FLAGWRD7
  CCS  A
  TC  WHIMPER -1  # YES.  DONT DO POODOO.  DO BAILOUT.

  TC  DOWNFLAG
  ADRES  STATEFLG

  TC  DOWNFLAG
  ADRES  REINTFLG

  TC  DOWNFLAG
  ADRES  NODOFLAG

  TC  BANKCALL
  CADR  MR.KLEAN
  TC  WHIMPER
```

The crucial BAILOUT code, which instructed the computer to drop less critical tasks and restart.

By placing their trust in painstakingly complex code and having faith in the engineer who vouched for the system in that critical moment, two astronauts, dozens of flight controllers, hundreds of technicians, and thousands of support personnel all came together to set sail on what President Kennedy had called "this new sea."[14]

None of the lunar modules flown in space returned to Earth. Some burned up on reentry, some were intentionally crashed into the lunar surface. One, from Apollo 10, entered solar orbit after its mission was complete. It's out there now, along with its Apollo Guidance Computer. The code embedded in the hardwired rope core memory, the work of hundreds of hands and thousands of minds over the course of a decade, is intact—and with it the hopes, dreams, voices, and fingerprints of the generations that built it. It is a monument every bit as eternal, and every bit as human, as boot prints in Moondust—a mute testimony to the tenacity and ambition of a species willing to attempt the impossible, step by step and line by line.

9

The Most Famous Comment in Unix History

"You Are Not Expected to Understand This"

David Cassel

It all started in 1975, with a routine chunk of code for the sixth edition of the Unix operating system, addressing something basic about how software gets run. Every application launched, even one running in the background, is considered a process. And back in the 1970s, before the dawn of personal computers, several different users would all be connected to one single central computer—meaning a system's processes included *all* the applications being run by *all* the users at any given time. These so-called time-sharing systems were constantly switching from one user's application to another's, quickly devoting a flicker of resources to whatever needed to be done next.

This also meant saving all the information needed for resuming later—like the numerical addresses in the memory where each application's data was being stored. (Called a "context," the information made it easy for the time-sharing systems to keep continually picking up right where they'd left off—a process known as "context switching.") This was all happening in an era when the whole operating system fit into just 9,000 lines of code. Twenty-five years later, by contrast, Microsoft's Windows XP operating system contained 45 million lines of code.[1]

Yet while that particular piece of Unix code may have turned obsolete decades ago, its fundamental concept is still very much in use today. It's what enables multitasking, the power to pause and resume programs with a single click. But what's most beloved about this code is a simple explanatory comment that accompanied it. The Unix operating system programmer at Bell Labs had written:

```
You are not expected to understand this.
```

Now, normally comments are pretty prosaic. Programmers are taught to include comments offering succinct explanations for what their code is accomplishing, as a way to help future readers of the code make sense of what's happening. Sometimes that future programmer may even be the person who originally wrote the code, now confronted with their own work years later and expected to make sense of it. My friend Brian Tanaka, a longtime programmer now working as a manager at Apple, once told me that he'd always thought of comments as "notes to *future* Brian."

But those seven words from 1975 would live on for decades.

Instead of warning off readers, the enigmatic comment intrigued them, becoming famous in its own right, and possibly for all the wrong reasons—an accidental icon, resonating with its unseen audience for the way it seemed to capture an all-too-familiar reality from every coder's world of intense technical complexity. At a 2016 tech conference, systems researcher Arun Thomas agreed that the comment had attained a kind of cult following among programmers.[2] "You started seeing people wearing sweatshirts, T-shirts, and baby onesies with '`You are not expected to understand this.`'" It was an expression of pride of technical mastery.

Thomas also shared another reason the comment became so popular: the code for the sixth edition of the Unix operating system soon became a standard educational resource, published in a book by John Lions, a professor at the University of New South Wales (along with Lions's own commentary). In fact, for many years to come, that book was "the *only* detailed kernel documentation available to anyone outside Bell Labs," according to a famous book on geek culture called *The New Hacker's Dictionary*.[3]

And then in 1979 the licensing terms for Unix were changed to allow only official licensees to study the code. So even after the code was replaced in later versions of Unix, that newer code was never made available in a book for college students to study. Eventually the owners of Unix clarified that even the book was supposed to be available only to official licensees. But this just meant that, as UCLA professor Christopher M. Kelty once put it, "Several generations of both academic computer scientists and students who went on to work for computer or software corporations were trained on photocopies of UNIX source code, with a whiff of toner and illicit circulation."[4]

The 1975 comment's fame became so persistent that nearly 30 years later, in 2004, Dennis Ritchie, one of the original programmers who wrote that Unix code, found himself creating a special web page just to share his own recollection of the real story behind it—if only to correct a common misperception.[5] "It's often quoted as a slur on the quantity or quality of the comments in the Bell Labs research releases of Unix," Ritchie wrote—as though the code was so horribly convoluted that no sane reader could be expected to understand it. ("Not an unfair observation in general, I fear, but in this case unjustified," Ritchie had added.)

Instead, he wrote, the original programmers simply felt that their code was addressing an obscure edge case, and "we tried to explain what was going on. '`You are not expected to understand this`' was intended as a remark in the spirit of 'This won't be on the exam,' rather than as an impudent challenge."

Ritchie also explained that ironically, "The real problem is that we didn't understand what was going on either." That is, they didn't understand the code *themselves*. It contained a bug that wasn't evident, since their hardware always returned the same value. But when Ritchie joined with his coworker Steve

Ken Thompson (sitting) and Dennis Ritchie working together in front of a mainframe computer. Thompson and Ritchie are credited with creating the Unix operating system at Bell Laboratories.

Johnson to try porting the Unix kernel onto a new machine, they'd discovered that "this code was indeed on the exam." For nearly a week their own bug thwarted any progress again and again, until they resignedly just rewrote the whole section from scratch.

Maybe the comment has remained so popular, despite the misunderstanding about what it meant, because it speaks to a general fondness and warm appreciation for those glimpses of personality that have sneaked into our computer code over the years. It's a feeling that can take many forms. Just take a look at the *Hacker's Dictionary* itself. It was originally just an ad hoc collection of geek culture and humor called "The Jargon

File," passed around online for more than 15 years, starting around 1976, by its fans on ARPANET (an early forerunner to the Internet of today). After years of being lovingly maintained by a community that had wanted to preserve it, it was respectfully collected up and published in 1991 by future open-source advocate Eric S. Raymond (under a new title, *The New Hackers Dictionary*). With its enlightening definitions and examples of geek wordplay, the book promises to describe "the language hackers use among themselves for fun, social communication, and technical debate." But what's just as significant is the book's implicit message: that there *is* a culture—a community with norms and values.

Of course, "`You are not expected to understand this`" isn't the only offhand comment to delight its future readers. In 2014, Silicon Valley's Computer History Museum announced they'd received permission from Microsoft to preserve the historic source code—and comments—for the original Word for Windows software released in 1990.[6] And one Microsoft programmer begins a comment by first dubbing their section of code a "gruesome hack." After a particularly technical flourish, another programmer added the comment: "Coded inline because we're god."[7] And there was also, not surprisingly, a large amount of profanity. (One site has actually created a graph showing how often various expletives appeared in the Linux kernel's source code over the last 30 years, with usage of the word "crap" recently attaining an all-time high.[8])

Some comments even become cherished reminders of a momentary glow of humanity in a world of unforgiving logic, where years are spent laboring around a language's requirement for perfect syntax. Programmers may humbly offer up a moment of self-deprecation, or warnings about straying too far into exotic and arcane solutions—and sometimes they even do both at the same time. When Unix pioneer Roger Faulkner passed away in 2016, a fellow programmer remembered one of Faulkner's cautionary self-deprecating comments:

```
/*
* This is a horrible kludge. It is vile. It is swill.
* If your code has to call this function then your code is the same.
*/
```
[9]

So there's an important meta-message hidden in these silent libraries of intellectual endeavors where moments for sharing are nonetheless baked into the process—that it's *people* who write programs. And, more importantly, that communities of *people* will maintain and preserve them. (The stereotype of a lone genius hacking away has become an outdated pop-culture artifact in today's world of massive coding collaborations.) Yet while code is written collectively, it's a community that's rarely seen—though often *felt.*

And every comment is an implicit acknowledgment of that community, and of all the careful caretakers who may someday be revisiting your code.

10

The Accidental Felon

Katie Hafner

Complex systems break in complex ways. That insight, from the prominent computer scientist Peter Neumann,[1] aptly describes the cascade of events that occurred across the nascent Internet one night more than three decades ago.

In the late 1980s, the Internet was still brand new, having just evolved from the original ARPANET. And though a handful of computer scientists were voicing concern about the havoc that could ensue were a rogue program to run loose in the network, no one was prepared to cope with the massive assault that took place on November 2, 1988.

At eight-thirty that evening, a 23-year-old Cornell University computer science graduate student named Robert Tappan Morris unleashed just such a rogue program into the network. His motivation was more curiosity than malevolence, his program intended as a harmless hack just to prove that it could be done.[2] The program was supposed to copy itself from computer to computer and simply take up residence in as many machines as possible, hiding in the background to escape detection by computer users. After he released it, he went to dinner, and in the hour he was gone, the program brought the Internet to its knees.

The incident grabbed the nation's attention as it painted in sharp relief, for the first time, the fragility of tens of thousands of interconnected computers. And it captured the nation's imagination because of the riveting father-son story it held: a father and son who belonged to a computer science elite, both obsessed with exploring the subtle intricacies of complex computers.

Robert Morris' father, Bob, was a brilliant computer scientist who had helped develop the Unix operating system,[3] which is the standard operating system for technical computing and was the target of his son's program. The elder Morris was the chief scientist for the National Security Agency's National Computer Security Center, the division of the agency that focused on computer security. For years, Bob Morris had tutored his son in methods for detecting and exploiting security weaknesses. Both father and son were well versed in a field in which a "game playing" mentality is essential: to make computers more secure, one must first be able to understand how to break into them.

The young Robert Morris, a quiet genius who found himself bored by many of his math and computer science classes, spent most of his undergrad time at Harvard's Aiken Computation Lab. Friends came to know him by his login: `rtm`. His senior thesis adviser recommended that he go to Cornell for his PhD, as it was a renowned center of computer science theory. If Robert was to be faulted for anything, it was his tendency to allow himself to be seduced by the machines themselves, at the expense of a theoretical understanding of the underlying science.[4]

And it was during his first semester at Cornell, with plenty of free time on his hands, that young Robert Morris wrote the program that would break the Internet.

Using several bugs he had found in the Unix source code, Morris designed the program to first steal passwords by reading the list of users on the target computer, then systematically running through their names, or permutations of their names, and a list of commonly used passwords. When successful in guessing one, the program then signed on to the computer and masqueraded as a legitimate user to copy itself to a remote machine.

Not only did Morris have no intention of crippling the Internet, but he had devised a mechanism for keeping the program from propagating too swiftly. Once the program had entered a computer, it would signal its arrival to the machine and ask

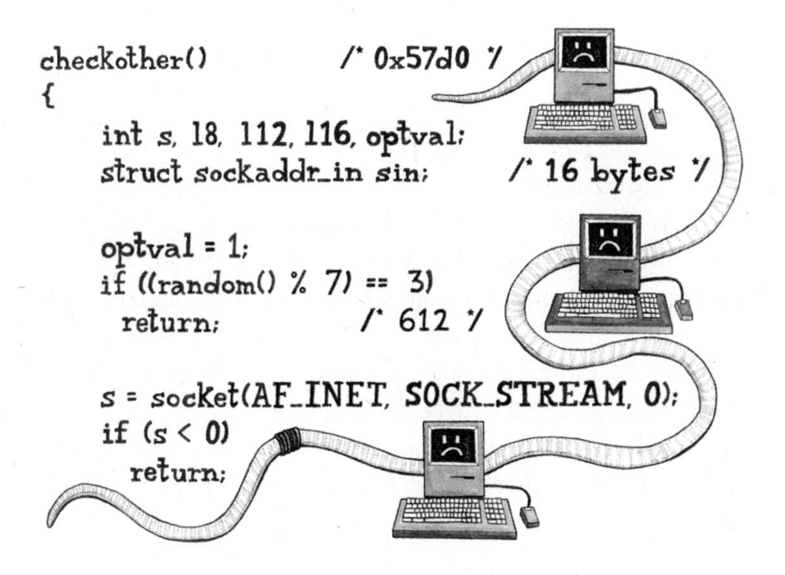

The code behind the 1988 Morris Worm, which crippled the nascent Internet.

whether that machine had already been invaded. If the answer was yes, the two copies would "talk" to each other and toss an electronic coin to decide which one should stop running.

But what if someone discovered the intrusion and tried to trick the incoming program into believing that it was already running on the machine it approached, essentially vaccinating the computer against the intrusion? Thinking like a chess player, Morris decided there would need to be a countermeasure against potential vaccines. His solution: randomization.[5] That is, one in *N* times, the program would enter a computer and command itself to run on the target machine regardless of the answer. Morris knew that the number he chose as *N* would bear directly on the rate of replication, but he wasn't sure what it should be. Ten? A thousand? Ten thousand?

The number the budding computer scientist ended up using—seven—turned out to be a singularly fatal miscalculation. The number should have been higher by a factor of a thousand or more, as the lower number resulted in dozens, even hundreds, of copies on each machine the program entered. The copies were like echoes bouncing back and forth off the walls of canyons.

By the time he returned from dinner to check on the program's progress, Morris realized, to his horror, that it had spread wildly out of control, slowing machines to a halt. Even when the program's attempts to get into a new computer were unsuccessful, its repeated knocks on the door were often enough to render the machine useless. Within hours the program had crashed thousands of computers at universities and research institutions throughout the United States.

Morris panicked and called a friend back at Harvard, even dictating a message for him to post from there:

```
A possible virus report:

There may be a virus loose on the internet.
Here is the gist of a message I got:

I'm sorry.

Here are some steps to prevent further transmission:

1) don't run finger or fix it to not overrun its stack when
   reading arguments.
2) recompile sendmail w/o DEBUG defined
3) don't run rexecd

Hope this helps, but more, I hope it is a hoax.
```
[6]

It wasn't a hoax. And it was too late. Computer managers stayed up through the night trying to fend off the virus as it hopped back and forth around the Internet, setting off havoc wherever it touched down. People were also frantically making other changes to their internal software to thwart future invaders, as they had no idea whether this was just the first in a series of planned attacks. At one of the computing facilities at UC Berkeley, someone made a sign that read "Center for Disease Control" and taped it to the door.

Soon after the incident, computer scientists decided that the program was technically a worm, not a virus. A virus remains dormant until the infected host file is activated, at which point it executes its code. A worm, on the hand, doesn't require the activation of its host file. Once a worm enters a system it can run, self-replicate, and propagate without a triggering event. The first worms were deployed on the ARPANET in the early 1970s. One was a benign program called Creeper, which resembled the Morris worm in that it copied itself from machine to machine. When it reached each new computer, it would display the message: "`I'm the creeper. Catch me if you can!`" As the Creeper story goes, a second programmer wrote another worm program that was designed to crawl through the ARPANET, killing all the creepers.[7]

Computer researchers started developing more useful worms, such as "town crier," a worm program that acted as a messenger, and "diagnostic" worms that patrolled the network looking for malfunctioning computers. There was even a "vampire" worm program, designed to run very complex programs late at night while the computer's human users slept. When the humans returned in the morning, the vampire program would go to sleep, waiting until the evening to return to its work.

But it was the Morris worm and its aftermath that permanently altered the culture of the Internet. Before the Morris worm, security was seen as more of a theoretical problem than a real one, and the network was like a rural town where everyone knows pretty much everyone else, where people leave their doors unlocked. The Morris worm changed all that. The incident raised fundamental questions about the security of the nation's computers and renewed debate over the question of who should be responsible for protecting the nation's nonmilitary computer systems.[8]

Computer scientists and systems administrators around the world were now on high alert, and any lingering naivete about the trustworthiness of those who used the Internet disappeared. Many computer security experts believed the Morris worm was an important and useful demonstration of the potential vulnerability of computers. Some even went so far as to argue that in heightening awareness of security issues, Robert Morris had done the field of computer security an enormous favor. Indeed, the Morris worm jump-started the field of computer security, creating demand for security experts in every quarter of the IT world. In the ensuing decades, every conceivable manner of malware came to plague the Internet, but the multibillion-dollar computer security industry can trace its roots to that signal event of 1988.

As for Morris himself, he was charged with a single felony count under the 1986 Computer Fraud and Abuse Act. It was the first charge under a provision of the law that made it illegal to gain unauthorized access to federal computers. A Syracuse jury convicted him. At the sentencing, a lenient judge spared him from prison and ordered him to do community service. Morris eventually finished his computer science PhD at Harvard, made a fortune in the computer industry, and became a tenured professor at MIT. In all these years, he hasn't spoken publicly about the worm and its legacy. The world has forgiven him his mistake, but we may never know whether he has forgiven himself.

11

Internet Relay Chat

From Fish-Slap to LOL

Susan C. Herring

If you were to spend 30 minutes on any Internet Relay Chat channel in the late 1990s, chances are good you'd see someone slap another person around a bit with a large trout.

So I observed when I started hanging out on Internet Relay Chat (IRC) as part of an ethnographic study I conducted in 1998. IRC is multiparticipant text chat that takes place in real time on forums or "channels" hosted on interconnected servers. Created by a student intern at the University of Oulu in Finland in 1988, IRC was flourishing at the time of my study, boasting hundreds of thousands of user-created channels, each name preceded by the # sign. For almost any topic you'd like to discuss or get help with, from `#altruism` to `#teensex`, there was an IRC channel that would serve your interests. Indeed, IRC was one of the earliest forms of social media, before social media existed as such. Although its popularity has been overtaken by web forums and social network sites since its peak in 2003, when there were more than 500,000 IRC servers worldwide,[1] as of 2021 IRC still existed on nearly 200,000 servers, and its chat protocol has been adapted for use on contemporary platforms such as Discord and Twitch.tv.[2]

Even if you've never heard of it, you probably use language in your online communication that originated in IRC. This includes that prototypical example of modern netspeak, `LOL` ("laugh out loud"). That expression (and its variants *lololol*, *lols*, *lulz*, etc.) traces its lineage back to the same source as the "fish-slap"—that is, constructions of the type:

```
* Susan slaps DrDoom around a bit with a large trout³
```

On IRC, such third-person, self-referential expressions are known as "actions," one of a set of line commands that allow IRC users to interact with the system and other users in a channel. Typing basic "slash" commands such as `/join` and `/nick` in the interface of an IRC client results in changes of state (such as joining a channel or creating a "nickname" or username). Similarly, one can type `/whois` (to see a user's profile information), `/ignore` (to prevent a user's text from appearing on one's screen), or `/part` (to leave a channel).

The trout-slap, in contrast, is an example of an *action command* produced by typing `/me` followed by a verb in the simple present tense plus any additional specifications. The output replaces `/me` with the user's nickname and displays a message in which the user appears to perform the action, like the example above of me slapping `DrDoom` with a large trout. On some IRC servers, the trout-slap became so popular that it got its own abbreviated command: `/slap`. Other abbreviated action commands include `/give`, `/laugh`, `/eye`, `/tag`, `/throw`, `/kiss`, `/hug`, and `/spank`.

Action commands typically describe purely *virtual* actions. IRC users, of course, are not literally slapping, spanking, kissing, or hugging others in a channel when they type these commands, nor does typing them result in any tangible change of state. At the same time, action commands create virtual realities that are not deniable. Thus, within the virtual world of IRC, `DrDoom` might protest at being slapped around with a large trout, but he couldn't plausibly deny that it happened. The command doesn't just describe the trout-slap, it *performs* it.

From these relatively obscure beginnings on the early Internet, virtual performatives spread, meme-like, throughout digitally mediated communication, morphing into forms that no longer require special commands or any knowledge of IRC to produce. `LOL` is a virtual performative; by typing it, you have effectively LOL'ed, whether you were physically laughing at that moment or not.

So how did we get from action commands to expressions such as LOL? I presume their evolution went something like this. Before there were abbreviated commands, a user who wanted to create a new virtual action on IRC had to type out most of the words. Consider this message:

```
* bonzo offers free weed to the first nine callers.
```

In order for it to appear, the user "bonzo" would have had to type:

```
/me offers free weed to the first nine callers.
```

However, since the /me command doesn't necessarily save keystrokes, some IRC users started to dispense with it, typing in virtual actions directly. Sometimes these direct expressions were set off by asterisks:[4]

```
<bonzo> *offers free weed to the first nine callers*
<Susan> *slaps DrDoom with a large trout*
<whitechellie> *happy sobs*
```

Other users left off the asterisks:

```
<DeANnA> dances for joy
<p2p> drops to tie his shoe
<DrDoom> chillin with the homies
```

Changes to the syntax of action commands took place, as well. As the examples above illustrate, direct virtual performatives typically omitted the grammatical subject, which was understood to be the users themselves, since their nick appeared by default (enclosed in angle brackets in the IRC client I used) at the beginning of each non-action message. Further, creative new uses of virtual performatives started to dispense with third-person present tense marking on the verb, producing uninflected forms such as `*gulp*`, `giggle`, and that now-classic `LOL`. Finally, the requirement for a verb was itself dispensed with. Instead, virtual performatives could be nouns, such as `*happy sobs*`, adjectives, such as `*reallyfuckingbored*`, or even expressive sounds, such as `*hrmph*`. Thus freed from the platform-specific, syntactic constraints of the `/me` command, performative expressions proliferated, spreading from IRC to other platforms. All the examples above are from my 1998 IRC logs, but similar examples can readily be found in private chat and on social media nowadays.

Part of the appeal of virtual performatives is that they are linguistically innovative. Not only do they riff creatively on the syntax of the IRC action command; they disregard the rules of ordinary (off-line) performative expressions. In English, for example, there are numerous constraints on what expressions can be used performatively and how they are expressed. The subject must normally be the first person ("I"),[5] as in "I apologize for being late" and "I promise to be good," which constitute an apology and a promise, respectively. Some performatives require institutional authority, such as "I sentence you to 10 years in prison without parole" (said by a judge). Only certain communication acts can be performative. You can't register a complaint, for example, just by saying "I complain about the new regulations." Moreover, verbs describing mental and emotional states can't be used performatively; you don't cause yourself to love cashews by uttering the sentence, "I love cashews." Finally, physical actions are excluded. There is no off-line context in which saying "I dance with joy" counts as an act of dancing (although it may count as a metaphorical expression of happiness).

In IRC and other text-based virtual worlds, in contrast, there is no difference between the uttering and the doing of an action. You can virtually "dance with joy" and "complain about the new regulations," as well as apologizing and promising, and you can adopt the role of judge and produce utterances like:

```
<DrDoom> *sentences you to 10 years in prison without parole*
```

As of this writing, IRC action commands—classics such as `/flirt`, `/hug`, `/insult`, and `/wave`, as well as newer commands such as `/bitcoin`, `/facepalm`, `/hookup`, and `/snopes`—are still used on some forum web chat and gaming chat servers, as well as in text chat in live streaming. Among these, the trout slap remains popular. Wikipedia has even expanded the command into the graphical realm, placing an image of a trout on a contributor's Talk page when the expression `{{trout}}` is added. This practice can be traced to the cyber-geeks familiar with IRC who formed Wikipedia's early user-base.[6]

It's doubtful that most social media users today know about IRC. Yet they still use IRC-like virtual performatives, such as this comment posted in a recent Facebook thread:

```
A newly minted assistant professor enters the chat
```

or this multimodal Twitter tweet:

```
*Sips tea ☕😯
```

This last example brings us to the most recent expansion of virtual performatives: emojis. Emojis are replacing many short-form performative utterances, such as 😂 (`lol`), 🤗 (hugs), 😉 (`winks`), 😘 (`kiss`), 😒 (`eyes warily`), 🤦 (`facepalm`), and 💃 (`dance for joy`). Moreover, they can combine to perform sequences of virtual actions, as in this tweet by a beauty influencer alongside a photo of herself at an airport:

```
Michelle Lewin
going places 💃👯
```

In the IRC of 1998, this would be expressed as:

```
<Michelle Lewin> going places *dances for joy*
*clinks champagne glass*
```

Thus, virtual performative constructions have come far from their origins in early multiparticipant text chat, all the while retaining their pragmatic force. Leaving me to wonder . . . shouldn't there really be a trout emoji?

12

Hyperlink

The Idea That Led to Another, and Another, and Another

Brian McCullough

Almost from the very beginning of computer science, you could split the field essentially in half: computers thinking for themselves (computation/calculation and artificial intelligence) and computers helping organize human thought. While the first gets more attention, the second is also challenging, asking whether computers can organize the infinity of human ideas—and the hyperlink comes very much from this second core idea in computer science.

The concept of linking ideas in some sort of functional way is one of those great conceits that kicked around for decades before finally settling into its perfect expression. Ted Nelson, one of the philosophical godfathers of information technology, coined the term and the idea of the "link" in his Project Xanadu back in the mid-1960s. Project Xanadu was a conceptual attempt at computer networking before the Internet was even born. But even Nelson was building off of ideas first expressed by another progenitor of modern computing, Vannevar Bush, back in his seminal 1945 *Atlantic* essay, "As We May Think."[1] Bush intuited a coming world where all of human knowledge would be accessible. But in such a future, the old ways of organizing thought via crude alphabetic or numeric indexing would be insufficient. In short, it was a problem of scale.

The human brain works by association instead of indexing, Bush figured, and thus if you could tie ideas together—linking them, if you will—it would be easier for humans to organize their thinking using the information computers were either storing or making available. Nelson built on this by proposing the "link" as a software concept to organize data by associations more amenable to human thought. This is the key insight of the link: it is a human brain–friendly software framework, a sort of fundamental, almost atomic-level schema for a human/machine interface.

Nelson's Project Xanadu spent decades lurching along as an ultimately quixotic attempt to build out a universal repository for knowledge. But all the while, Nelson held fast to Bush's idea of the link as a cross-referencing concept either for specific text strings (thoughts) or documents (fully realized ideas). Independently, Douglas Engelbart implemented the link concept for items within a single document when he was working to develop modern computer interfaces.

From there, the direct lineage of the hyperlink gets muddled. Hypertext, the formatting of text for computer displays and electronic devices, came to prominence beginning in the 1960s (thanks to Nelson again), and the term "hyperspace" arose as a mental framework to imagine the sort of ethereal realm where these interlinked documents existed. Thus, the highlighted and underlined string of text used in the HyperTIES system,[2] one of the earliest stabs at creating a "browser" for information retrieval (and which Tim Berners-Lee citied in his original proposal for the World Wide Web),[3] naturally gained the sobriquet "hyperlink." In the mid-1980s, the database program HyperCard was released for the Apple Macintosh and popularized the linking between various pages within a document as well as linking out to other documents.

And then came the graphical user interface, with its clickable icons. All of the complexity of computing hidden away behind a friendly abstraction.

An often-overlooked element of early computing was the "tyranny" of command line. To make a computer do what you wanted, you *already had to know how* to tell the computer what to do. Not only did you have to understand syntax and protocols and instructions, you had to have a mental map of a computer's file structure. But then, suddenly, serendipitously, the graphical user interface subsumed all of this arcana behind colorful icons and graphics, and you could tell a computer what to do simply by clicking or dragging and dropping.

This user interface revolution was also long needed on the Internet, which for its first 20 years struggled under the twin burdens of the command line and file structures. To go somewhere on the Internet you needed to know where to go, often a string of numbers. Even when the Internet Protocol system was made more humane so that `182.06.254.1` (a hypothetical example) could be represented by something more manageable like "Amazon.com," you still had to have a list of all the places you might want to go. And if you wanted a specific file, say "`apple.com/x/file-source/25473s.xyz`," again, you had to type your directions no matter how inscrutable they were.

When Tim Berners-Lee announced the World Wide Web as a concept in 1989, he drew from this entire mixed lineage of the link and the user-friendly paradigm of the click, to create the key concept of the hyperlink.

And thank God he did.

The problem for the Internet has always been scale. If you connect all the computers in the world together, then you (in theory) have connected all of humanity's collective knowledge. But at that point, the needle in the haystack analogy is insufficient to describe the problem by orders of magnitude. The hyperlink dispensed with any notions of "organizing" global information. The World Wide Web of associations would do the organizing organically and intuitively. Sometimes people like to talk fancifully about printing up every entry in Wikipedia, stacking them alphabetically, speculating on how high that stack of pages might go. (Artist Michael Mandiberg actually did this in an installation first exhibited in 2015.) But this misses the point. The idea of one idea leading to another, and then another, and then another, click-click-click, does not just give us the ecstatic intellectual experience of "going down a rabbit hole"—it is the *very mechanism* that allows organized thought in a digital age to happen.

The great triumph of the Internet Era is that we connected all the computers in the world together, and thereby set out to map and encode all of human thought. The fact that humans can access their own collected knowledge is because the hyperlink strings it all together in the same way our brains are wired. The hyperlink brings software concepts to text and the structure of the human brain to digital data. It is a conceptual melding in both directions: making the act of computing more experientially "brain-like" in terms of user experience, and making ideas more "computer friendly" in terms of user interface. I don't have to know exactly *where* in hyperspace a given idea or document is, I only have to know that, via a hyperlink, someone has already pointed me in that direction. To journey there, I only have to click.

This points to one more way the hyperlink, at least as ultimately manifested by Berners-Lee's World Wide Web, makes the modern Internet possible. The data I am seeking can be stored in any number of incompatible systems. When Berners-Lee was attempting to first explain why the Web would be useful to anyone, he referenced the confusing status quo of networked computing:

> I found it frustrating that in those days, there was different information on different computers, but you had to log on to different computers to get at it. Also, sometimes you had to learn a different program on each computer. So finding out how things worked was really difficult. Often it was just easier to go and ask people when they were having coffee.[4]

So the hyperlink also solved the problem of interoperability on a fundamental level. If your data is formatted in HTML, or any number of the modern Internet standards, most every computer system in the world can read it.

Of course, the hyperlink is not perfect. For all his protestations that he eventually intended to make it go both ways, Berners-Lee's classic hyperlink is unidirectional. If I link from my blog to yours, a third party can see the direction of intent from me to you. But what happens if they were to stumble upon your blog first? The value of the citation does not flow backwards. And believe me, there is value in both directions. You only have to look at the market cap of a little company called Google to realize this. The other great organizing principle of modern digital thought, "search," was only made possible because Google set out to reverse-engineer hyperlinking retroactively. Furthermore, the idea that the "value" of links could be tied to actual value (i.e., money) is something that blockchains are only now beginning to solve, along with the problem of "link rot"—the unfortunate reality that the permanence of a hyperlink depends on the content to which it links continuing to exist.

But the hyperlink as the atomic unit of the digital age, both for software and for human thought, is probably best proven by the fact that it likely made the concepts of social media manifest for everyday—even nontechnical—people. After all, what is a like but a link? If I "follow" you, is that not a form of citation? If the original Internet revolution was about linking all of the computers and then the ideas in the world together, then the social graph was merely phase three: linking all the people in the world *and* their affinities together, in the same way the hyperlink mapped ideas.

13
JPEG
The Unsung Hero in the Digital Revolution

Hany Farid

In 2008, I received an unexpected call from a detective in Scotland. Beneath the detective's Scottish accent, I heard a strong sense of urgency and seriousness. A dozen men stood accused of abusing young children and distributing images of the abuse. At the center of the complex, multiyear case was a series of images of unknown origin. The detective asked whether I could link the photographs to one of a handful of cameras that had been seized in the investigation. I could, thanks to the fact that the images were saved as JPEGs.

The JPEG image format is the standard compression scheme for digital cameras. Compression schemes allow for the trade-off between image file size and image quality. A highly compressed image requires relatively little memory for storage and transmission but may have noticeable distortions. A less compressed image will have greater fidelity to the original but requires more memory and bandwidth. The ubiquitous JPEG standard was established in 1992 based on a compression scheme proposed in 1972, which was itself based on basic mathematics dating back to 1882.

At the point of recording, a digital image is made up of an array of picture elements, or pixels. Each pixel is itself composed of three numbers corresponding to the primary red, green, and blue colors (RGB). An uncompressed, modest-sized, 1000 × 1000 pixel RGB image consists of one million pixels and requires approximately three megabytes of memory to store on a camera or computer. These days, digital cameras record images with tens of millions of pixels which, left uncompressed, each require more than ten times this amount of memory.

In the early days of the Internet, digital media was highly compressed with relatively low quality. But at least we could share audio, images, and video. As bandwidth, as well as computer and device memory, increased, it became easier to store and transmit increasingly higher-quality content. Without data compression, however, it would have been impossible to record, store, and share uncompressed content at the scale we do today: nearly two billion images per day, and, on YouTube alone, 500 hours of video every minute. The untold hero in this digital landscape is French mathematician and physicist Jean-Baptiste Joseph Fourier (1768–1830). In his seminal 1822 work on heat flow, Fourier made a claim that 100 years later would play a critical role in the digital revolution: Fourier claimed that any function can be expressed as a sum of multiple sinusoids[1] (a sinusoid, or sine function, oscillates in value smoothly through peaks and troughs).

The shape of the elegant sinusoid—sin(ω)—can be described by spinning a line around a circle and measuring the vertical distance between the circle's center and the line's tip. The speed with which the line spins around the circle defines the sinusoid's *frequency*—the number of oscillations per second; the length of the line defines the *amplitude*—the height of the oscillations; and the starting position of the line defines the sinusoid's *phase*—the relative position of the oscillations. A high-frequency sound like a squeak, for example, has many rapid oscillations in air pressure per second, while a low-frequency sound, such as a rumble, has fewer, slower oscillations per second. Turn the radio volume up or down, and the sound's amplitude increases or decreases.

There is a visual analogue to a sound's sinusoidal representation. A high-frequency visual pattern has many abrupt changes in appearance across space—picture the texture of grass—while a low-frequency visual pattern has only gradual changes across space—picture a cloud pattern. Similar to volume, amplitude corresponds to the brightness of the visual pattern.

As we will see next, Fourier's insights into the power of the sinusoid to represent signals and patterns laid the groundwork for an efficient way to digitally represent, store, and transmit audio, image, and video, in turn revolutionizing the power and reach of the Internet.

That brings us to JPEG compression. The simplest way to compress an image is to throw away pixels. Starting with a 1000 × 1000 pixel image, for example, throwing away every other pixel results in a 500 × 500 pixel image with a total of only 250,000 pixels as compared with the original 1,000,000 pixels, for a savings of 4×. This, however, is highly undesirable. Why should we build high-resolution cameras, capable of recording high-fidelity images, only to reduce the image resolution immediately after recording because we can't store or transmit the images?

We seek, therefore, to compress an image to reduce memory and transmission costs, while retaining resolution and visual quality.

The digital-camera revolution was kick-started in 1969—the same year as the Apollo moon landing—when Willard Boyle and George Smith invented the charge-coupled device (CCD) for electronically recording and storing light. Around this same time, researchers were already considering how to best compress digital images. The Karhunen-Loeve transform (KLT) emerged as the best way to compress digital data. This transform, however, was computationally costly, leading Nasir Ahmed in 1972 to develop the Discrete Cosine Transform (DCT),[2] itself inspired by Fourier's insights into the power of sinusoidal representations.[3] The DCT quickly emerged as an effective and efficient way to compress digital images and eventually was adopted by the Joint Photographic Experts Group that, in 1992, established the JPEG compression standard.

The JPEG compression standard was designed to take advantage of the human visual system's differential sensitivity to various forms of visual information. This compression scheme attempts to preserve the image information to which we are most sensitive while discarding information we are unlikely to notice. For example, we are more sensitive to luminance contrast—a change from light to dark—than to color contrast—a change from red to green. Consequently, JPEG compression preserves more information about luminance than about color. JPEG compression also treats frequencies differently. Humans are more sensitive to low frequencies (cloud pattern) than to high frequencies (grass texture), and, accordingly, JPEG compression preserves more information about low frequencies than about high frequencies. (Audio [MP3] and video [MPEG] compression operate on the same concept of transforming the original data to a frequency-based representation and differentially compressing based on human sensitivity.)

While there are many details in the complete JPEG compression scheme, the heart of this compression relies on representing visual patterns using sinusoids (or, more precisely, a phase-shifted version of the sinusoid, the cosine) and removing content to which the human visual system is less sensitive. The heart of the DCT is this variation of the Fourier transform:

$$F(x,y) = a_{x,y} \sum_{u=0}^{7} \sum_{v=0}^{7} f(u,v) \cos\left(\frac{(2u+1)x\pi}{16}\right) \cos\left(\frac{(2v+1)y\pi}{16}\right)$$

The DCT transforms each 8×8 pixel image block (f) to a frequency-based representation (F), allowing for differential compression to different frequencies. This compression is achieved by rounding small values in the high-frequency range to 0 (which can then be efficiently represented in the final JPEG file), while preserving the more visually salient low frequencies.

Although JPEG compression allows for fine control over the compression of each frequency across each luminance/color

```
# input:   B: 8x8 image-pixel block
#          Q: 8x8 quantization (integer-valued)
# output:  D: 8x8 block-DCT
def dct2(B,Q):
    D = np.zeros((8,8)) # initialize
    y,x = np.meshgrid(np.arange(1,9,1), np.arange(1,9,1))
    for i in range(1,9):
        for j in range(1,9):
            ai = np.sqrt(1/8) if i == 1 else np.sqrt(2/8)
            aj = np.sqrt(1/8) if j == 1 else np.sqrt(2/8)
            D[i-1,j-1] = ai * aj * np.sum( np.sum(B * np.cos(np.pi*(2*x-1)*(i-1)/16) *
                                                      np.cos(np.pi*(2*y-1)*(j-1)/16)) )
    return( np.array( D/Q, dtype=int) ) # quantize
```

A modern JPEG encoder is highly optimized to quickly compress and decompress images. This Python code snippet implements a nonoptimized version of the basics of a JPEG encoder, consisting of the DCT transform and DCT quantization.

channel, all standard photo-editing and coding libraries synthesize these compression parameters into a single setting ranging from high-compression/low-quality to low-compression/high-quality. For example, I compressed an 8-megapixel image across an entire compression range yielding, at one end of the compression/quality spectrum, a 0.2 MB file size, and at the other end, 5.3 MB—the uncompressed image came in at a whopping 25.8 MB.

Beyond playing its part in the digital and Internet revolution, JPEG compression has played a critical and unintentional role in the forensic analysis of digital images. Which brings me back to that call from Scotland.

My forensic analysis required a two-step process, the first of which leveraged distinct JPEG compression settings that vary across devices and software. Most notably, the luminance/color and frequency-specific quantization values vary as a result of different compression and visual distortion tolerances for low-, medium-, and high-end cameras. Because these tolerances are constantly being refined, even successive releases of the same camera may use different compression settings.[4] These variations allowed me to identify the make/model used to record the images in question. A secondary analysis allowed me to uniquely identify the camera based on subtle imperfections in the underlying camera sensor.[5] This forensic analysis, along with some exceptional investigatory work by Scottish police, led to the conviction of a dozen child predators.

14

The Viral Internet Image You've Never Seen

Lily Hay Newman

It's likely that the most downloaded image ever isn't the Mona Lisa, the Google logo, or the first page of the King James Bible. In fact, you've never seen it, even though your browser probably requests it from servers every day. That's because it's not a famous photo or illustration—it's a single transparent pixel[1] that's used by all sorts of entities to silently gather data about you and your web activity. These minute pixels are scattered across websites and embedded in Emails. Simply by existing, they gather identifying data like your IP address, what browser you're using, and which operating system you're running. And they're everywhere. This is the unseen power, and menace, of the tracking pixel—a relic of the '90s Internet that has played an outsized role in the twenty-first-century digital marketing machine.

Also known as "1 × 1 pixels," "web bugs," and "invisible GIFs," among other names, transparent 1 × 1 pixels were originally used as visual spacers in early web page layouts. Before the 1996 debut of Cascading Style Sheets (or CSS, the web programming language used with HTML for content layout), web developers relied on an array of hacks and workarounds to create the visual elements they wanted and get everything aligned. Tiny transparent images could sit next to text boxes or visible images to make room between components or smooth out any awkward formatting—like the blank type spacers used in physical printing.

To a web server, though, a transparent single-pixel image is just like any other. When a user navigates to a web page or opens

<IMG SRC="TRACKING" WIDTH="0" HEIGHT="0">

An example of code for an Email tracking pixel.

an Email, their browser or mail client sends a request for the images on that page or in that message and for details about how to display them. When the browser initiates that interaction, its request includes information like the time, the device the browser is running on, and its IP address. That simple, basic exchange can then give marketers, advertising networks, or any organization information about who opens their Emails and visits their web pages. These details can be combined with usage data from other digital tracking tools, especially cookies, to produce a profile of a user's activity, interests, preferences, and even physical location.

"The user would never know that the pixel is there and it's able to do things that the user would never expect or want," says Jason Kint, CEO of the digital publishing trade organization Digital Content Next.[2]

In the mid-1990s, marketing firms on the vanguard of digital ad tracking realized that 1 × 1 transparent pixels provided an easy, inconspicuous way to get more granular information about what their target audiences were up to online. And privacy advocates were close behind on warning about the implications and dangers of such behavior.

"Clearly Web Bugs are controversial," privacy researcher Richard M. Smith wrote in a 1999 Electronic Frontier Foundation guide titled The Web Bug FAQ.[3] "Because they allow people to be monitored, when they don't expect it, they certainly can be very upsetting. For example, most people will likely be troubled to learn that an outsider is tracking when they read Email."

Early Web users certainly wouldn't have expected that invisible beacons were tracking their online activity. Even cybersecurity experts who understood the potential implications of the technology struggled with tracking pixels' pervasive ubiquity.

"I remember thinking about them in email at least fifteen years ago or more," says longtime digital security and privacy researcher Katie Moussouris. She recalls "tracking pixels just driving up my anxiety about opening email and being annoyed at it being used by marketing spammers."[4]

Even for web users with technical prowess, like Moussouris, the most sinister aspect of tracking pixels has been their profound simplicity and inveterate reach. To mitigate tracking pixels yourself, you generally have to use browser plugins that can break website features, disable all image requests on a site, or only view Emails in plaintext.

"Users are able to turn off cookies or even remove individual cookies, but it's not so easy with tracking pixels, which are mechanical elements of tracking scripts, embedded on websites," says independent privacy researcher and consultant Lukasz Olejnik.[5]

And the tracking pixel was not a '90s flash in the pan like one-hour delivery service Kozmo. Three decades after their creation, you regularly encounter tracking pixels across the Internet. But you might not realize it, even though web users are more aware than ever of digital tracking and the forces of targeted advertising.

"The scale and techniques are perhaps more advanced today, but the fundamentals are the same," says former Federal Trade Commission chief technologist and senior White House advisor Ashkan Soltani, who gave Congressional testimony about the state of online tracking and digital privacy in 2011 and 2018. "The web was created in this fashion that leaves a digital trail everywhere you go and permitted third parties to observe that trail and link it to you. That didn't need to be the case, but that's what happened."[6]

One descendant of the tracking pixel is Facebook's "Like" button, which is embedded in websites across the Internet. The buttons are essentially tracking pixels that detect cookies and use other identifiers to collect details about what you're up to online and fuel Facebook's ad targeting empire. The company also offers an invisible option known as a "Facebook Pixel" that businesses can embed in their websites to track new and existing customers who visit the page, target ads, and gather data on what happens after visitors see a given ad.

"This isn't just a tinfoil hat paranoid thing," security engineer and independent researcher Kenneth White says of tracking pixels. "This is one of the fundamental gears of a machine that's at the heart of the global economy."[7]

Decades after privacy advocates first warned about tracking pixels and the rise of targeted marketing, mainstream companies have recently started expanding their offerings to give regular people more control. Apple's AppTrackingTransparency framework,[8] for example, which launched for users in April 2021, requires that apps ask for permission to track users across multiple different services. The goal is to give users more choice and let them know how often apps are tracking them across seemingly disparate services, like those that incorporate Facebook Pixel.

When it comes to blocking tracking pixels specifically, there have long been some options available if you're willing to make aesthetic and functional sacrifices. Google's Gmail, for example, offers a setting to "Ask before displaying external images,"[9] blocking image loading entirely unless you individually approve specific content. But now these types of tools are moving away from broad, heavy-handed blocking in favor of a more tailored and targeted approach.

Apple announced in June 2021 that its iOS 15, iPadOS 15, and macOS Monterey operating systems would have a specific feature in their "Mail" apps that stops Email web bugs in their tracks. The "Protect Mail Activity" feature loads your Emails through a special set of proxy servers to mask your IP address and details such as the time that you opened an Email, so trackers aren't getting accurate or helpful information. The privacy-focused web services company DuckDuckGo announced a similar feature, Email Protection, in July 2021.

"We're seeing some people and companies reevaluating assumptions, because they're realizing that the free love Internet of the 1980s and '90s had unintended consequences in a globally connected, financially motivated ecosystem," says Soltani, who became executive director of the California Privacy Protection Agency, "and one of the assumptions they're looking at is, should we allow invisible third parties to profile us and track us on the Web?"

For the average web user who's been subjected to marketing surveillance and targeted advertising for decades, this industry introspection may feel almost laughably overdue. Small gestures from tech giants often end up highlighting bigger, more comprehensive transformations that are not taking place. For example, Google's recent efforts to kill third-party cookies in Chrome[10] have been repeatedly delayed and revised. The company proposed a plan in February 2021 that involved an alternative in which Google alone would retain the ability to implement anonymized, cookie-like activity tracking.[11] Almost a year later, amid backlash from both marketers and privacy advocates, the company debuted a new proposal to offer "interest-based advertising" based on broad categories that your browsing fits into on a given week, like "Fitness" or "Travel & Transportation."[12] And in February 2022, Google also committed to phasing out inveterate and invisible cross-app tracking, following Apple's move in 2021.[13] After decades of maturation, the targeted ad industry certainly won't be dismantled in a couple of years. But if something as innocuous as an invisible pixel can fuel a marketing revolution, perhaps a privacy revolution can come from something equally humble.

15

The Pop-Up Ad

The Code That Made the Internet Worse

Ethan Zuckerman

Sometime around 1997, I wrote a line of JavaScript code that made the world a measurably worse place. The line of code looked something like this:

```
window.open('http://tripod.com/navbar.html'
"width=200, height=400 toolbar=no, scrollbars=no,resizable=no
target=_top");
```

This line of code was inserted into the top of every personal home page we served at Tripod.com, one of the pioneers of "user-generated content," the not-especially-radical idea that the Web would be built by ordinary users, not by professional media companies. When it loaded in a web browser, the code opened both the personal homepage the viewer had wanted to see, and another window, which included Tripod branding, tools for navigating between different user-generated pages, and a 200 × 200 pixel ad. It was likely the Web's first pop-up ad.

The pop-up ad was my inelegant solution to a thorny problem the contemporary Internet still faces: How can advertising support user-generated content? Advertisers were excited to reach the tens of millions of viewers who came to view web pages hosted on Tripod, but they often didn't want to be too closely associated with the contents of those pages, which could be crudely produced, in questionable taste, and sometimes ob-

The code that created the scourge of the Internet.

scene. The pop-up specifically came about after an auto company complained about their ad appearing on a personal homepage about anal sex. My boss asked me to find a way to sell ads while ensuring brand managers wouldn't send us screen shots of their precious brands juxtaposed with offensive content. My slapdash solution? Put the ad in a different window than the content. Presto! Plausible deniability!

A few weeks later, our leading competitor, GeoCities, copied our code and used it to launch a similar window on their pages. (This didn't exactly require sophisticated corporate espionage—JavaScript code like this was visible on the top of a web page's source code, which could easily be viewed within a web browser. Many JavaScript coders learned their trade primarily from reading web pages they found online.) Not long after, I saw a pop-up window appear with only an ad in it. Then came pop-ups that moved when you tried to close them. Pop-unders that loaded below a web page and revealed themselves when you closed it. Pop-ups that spawned other pop-ups in a cascade of unwanted advertising that made you regret you'd heard of the Internet in the first place.

Within a few years, the option to block pop-up windows was built into most web browsers, and advertisers moved onto other ways to seize an unwitting user's attention: ads that began playing music until you clicked on them, videos that began playing as the rest of the page loaded. Pop-up ads were an artifact of the early Web, like the <BLINK> tag that made text strobe until designers realized it was a terrible idea and stopped using it.

I had largely forgotten my role in polluting the World Wide Web until August 2014 when I wrote an essay for *The Atlantic* arguing that the business model of offering services at no monetary cost in exchange for surveilling users and monetizing our attention was the "original sin" of the Web. It was a long and nuanced essay with literary pretensions and engaged with emerging critiques of the Web, like Shoshana Zuboff's idea of "surveillance capitalism," while prescribing a Web supported by subscriptions and micropayments as an alternative.[1]

My editor identified the single sentence of the article she knew readers would latch onto: my apology for unleashing the pop-up ad as part of a desperate attempt to make advertising work on user-generated content. She interviewed me about the pop-up and ran the interview as a 300-word story that generated ten times as much traffic as my 4,000-word essay.[2]

Within 48 hours, late-night TV hosts were cracking jokes about my apology. "The guy who created the pop-up ad says he's sorry. He also says you can save on auto insurance with GEICO!" My inbox filled with a variety of invective—to this day, I know a content-hungry news website somewhere has published a new version of my pop-up "confession" because a random teenager has reached out to express his desire to travel through time and kill me instead of Hitler.

Death wishes aside, the critique I more often receive today is the opinion that I'm taking too much blame—and too much credit—in accepting responsibility for the pop-up ad. They accurately point out that someone else would surely have created the ads had I not unleashed them. Why harp on my minor role in Internet history? Is this not just a desperate plea for attention, a reminder of my presence in the Internet industry two dozen years ago?

I recently had the opportunity during a virtual conference to ask David Bohnet, the CEO of Tripod's competitor, GeoCities, whether he had any regrets about the role our firms had in bringing about the ills of the contemporary Web by linking user-created content to surveillant advertising. He forcefully disclaimed any responsibility, noting that the phenomenon of targeted advertising preceded the advent of the Web. He noted with pride that GeoCities' "neighborhoods," where individuals could opt into communities of interest, had helped make an ad-supported Web viable. When fans of country music built their virtual homes in the "Nashville" neighborhood, they signaled that interest to advertisers, perhaps allowing the advertiser to intuit demographics as well. Advertising based on these stated intentions could be less invasive than third-party cookies and other advertising technologies that follow us around the Web today. The conference's moderator, knowing my history, chided me for shouldering too much blame for the Web's contemporary ills and trying to share that burden with my former competitor.

I appreciated David's honesty, but found myself pondering a deeper question: Who is responsible for the current dysfunctions around social media? In the wake of Russian interference in the 2016 US election, the scandal around Cambridge Analytica, and the coordination of a mob that invaded the US Capitol on January 6, 2021, it's not hard to conclude that something is amiss in social media and its effects on the general public. Mark Zuckerberg often positions himself as a single, easily hateable figure to absorb this blame. But an accurate attribution of blame is more complicated.

Zuboff describes "surveillance capitalism" as a new economic system in which each action we take, each purchase we make, each interaction we have becomes data to be used to predict our future behavior. This is not a business model, Zuboff argues, so much as a new economic paradigm—it invades all businesses, whether they are advertising-driven or not, using the logic of "big data" to turn behavior and reality itself into source material for processing and prediction.[3]

Facing a shift of this magnitude, who do we credit or blame? Google is Zuboff's particular bête noire, but a case could be made to blame a generation of venture capitalists who rewarded rapid user growth above all other factors. We could blame regulators who've allowed a massive new economy of data sharing and sales to emerge almost without oversight. Or computer science professors who have taught programming as a pragmatic technical skill with little thought toward ethics and values. Or countless engineers who made bad design choices in trying to navigate a set of business requirements they didn't think to question. People like me.

The problem with refusing to accept personal responsibility for making the Web a more terrible place is that it releases you from the responsibility to make it better. If the emergence of surveillance capitalism is diffuse and multifocal, it becomes a force of nature, unstoppable and incomprehensible. If it's the sum of small decisions made by self-interested actors, it's understandable and possibly reversable.

My work these days is as an academic, teaching at the University of Massachusetts Amherst and leading a research group focused on alternative economic and governance models for social media. In other words, I'm working to build a version of social media that doesn't watch you, predict your behavior, and target ads to you. It turns out there's a variety of ways to do this, including subscriptions, micropayments, cryptographic tokens, sponsorship, non-surveillant advertising, or by thinking of digital public spaces as public goods, supported by taxpayer funds or volunteer efforts. Some of these models might create the next venture capital unicorn, but most merely enable new, much smaller communities that might behave very differently than Facebook or YouTube.

Personally, I believe that social media should look more like public media, designed not to turn a profit but to address issues and conversations that help us live together in a democracy. My team is building small, single-purpose social networks designed to host moderated, civil conversations about local issues. Our goal is to help people reimagine social media as a space for constructive interaction, rather than the toxic stew it so often becomes. We are also building tools that give users more control over their social networks, allowing each of us to decide what content we'd like to see more and less of.[4]

It turns out that the hardest part of inventing alternative futures for the Internet is giving yourself permission to imagine something radically different. My colleagues and I are imagining a future where social media is a key part of a healthy media ecosystem, helping us find allies and common ground, and argue productively with those we disagree with. But we also need to imagine responsibly. A great first step in stretching your imagination is accepting responsibility for your role in making the Web what it is today, even if your only responsibility has been not demanding something better.

16

Wear This Code, Go to Jail

James Grimmelmann

WARNING
This shirt is classified as a munition and
may not be exported from the United
States, or shown to a foreign national

This was the stark warning printed in high-impact white-on-black type on the front of the "RSA T-shirt." Beneath it were four lines of what looked like the output of a botched print job:

```
#!/bin/perl -s—-export-a-crypto-system-sig -RSA-3-lines-PERL
$m=unpack(H.$w,$m."\0"x$w),$_=`echo "16do$w 2+4Oi0$d*-^1[d2%Sa
2/d0<X+d*La1=z\U$n%0]SX$k"[$m*]\EszlXx++p|dc`,s/^.|\W//g,print
pack('H*',$_)while read(STDIN,$m,($w=2*$d-1+length($n)&~1)/2)
```

This dense little block was actually the code for the RSA encryption algorithm, implemented in the Perl programming language. It was the position of the United States government that it was illegal to export this code without a license—making the T-shirt bearing it into a regulated "munition" that could not be given or sold to foreigners.

No one was ever prosecuted for letting America's enemies wear the "RSA T-shirt," but if they had been, it would have proved Adam Back's point about the absurdity of export-control regulations in a digital age.[1] Back and his fellow cypherpunks believed in digital freedom of speech, freedom to tinker, and freedom from surveillance. They created the shirt to protest ap-

The RSA T-shirt, a regulated "munition" that could not be exported from the United States.

plying export-control laws to software. It was their contribution to one of the most memorable skirmishes of the crypto wars of the 1990s, a kind of wearable argument for the right to encrypt.

The idea behind export controls is straightforward: some technologies are too dangerous to let fall into the hands of a nation's enemies. If you're in the United States, you can't sell howitzers, anthrax, or space lasers abroad without government permission.[2] They're considered "munitions" by law, and the State Department will ask you some pointed questions about who you're selling them to. It's not just weapons per se on the export-control lists, either. High-volume fermenters and centrifuges need licenses, too, because they can be used to make and isolate viruses.[3]

The export-control rules have always been an awkward fit for digital technology. The relentless march of Moore's law means that last year's military-grade electronics are often next year's must-have Christmas present. For a time, the PlayStation 2 technically qualified as a "supercomputer" subject to export control, because it was as powerful as the multimillion-dollar systems previously used to do the intensive computations required to design nuclear weapons.[4]

Encryption ended up on the export-control lists because of its ability to let the dreaded bad guys talk in secret. It made a certain kind of sense when encryption was a feature of specific physical devices, like battlefield radios that scrambled their signals. But as encryption became a feature implemented in software, any computer could become an encryption device. To keep up, governments tried to apply their export laws to software too, not just the computers it ran on.

Enter the crypto geeks. Since ancient times, cryptography had relied on "symmetric-key" algorithms.[5] The sender and the receiver both know a key, which has to be kept secret from everyone else in the world. The sender uses the key to encode a message; the receiver uses it to decode the message. The difficulty of setting up a safe exchange of keys made symmetric-key cryptography hard to use, giving government snoops and codebreakers a leg up on their surveillance targets.

But in the 1970s, cryptographers devised a new class of asymmetric "public-key" algorithms that blew up the key-distribution problem.[6] Now, the sender could encode a message using a widely known public key, while the receiver decrypted it with a private key known only to her.

The first practical public-key algorithm was created in 1977 by three MIT researchers, Ron Rivest, Adi Shamir, and Leonard Adelman, and named "RSA" for their initials. The idea behind RSA is that, as far as anyone knows, it is much easier to multiply two numbers than to factor their product to determine the original numbers. Simplifying a bit, the receiver's secret key is a pair of very large prime numbers; the public key, suitable for use by any sender, is their product. Messages are encrypted by writing them as numbers, raising them to a specified power, and then dividing them by the public key and taking the remainder. Only the receiver, who knows the factors of the public key, is able to reverse the process and obtain the original message.

Inspired by RSA, many other researchers developed their own public-key algorithms and created new applications for them. Want to sign a document securely? Encode it with your own private key; anyone can decode it with the public key and know that only you could have encoded it. The 1980s were a time of ferment and rapid progress in cryptography.

The pace of crypto research, however, ran headlong into the export-control laws when Daniel Bernstein, a graduate student at the University of California, Berkeley, developed an encryption algorithm charmingly named Snuffle. In 1992, the State Department told Bernstein that Snuffle fell within the definition of a "defense article" in the export-control regulations. He would need a license to export Snuffle—which could include mailing copies of his paper abroad or presenting it at conferences where there were foreign nationals present. He couldn't even post Snuffle's source code online without permission, because there was no way to prevent foreigners from downloading it.

Bernstein sued the State Department with the help of an all-star team of digital civil-liberties lawyers and asked the court to declare that he had right to share his research publicly.[7] While his case was pending, Peter Junger, a law professor at Case Western, filed his own suit against the export-control regulations seeking the right to post encryption software on his website so he could teach about it in his course on computers and the law.[8]

Bernstein's and Junger's suits energized an increasingly activist community of technologists. A loose group of "cypherpunks" saw strong encryption and anonymity as human rights and defenses against government abuse. As a community, they resisted other attempts to control software and encryption. Most famously, they fought back in 1993 and 1994 when the Clinton administration tried to mandate that telephone encryption use a standard called Clipper that included a government backdoor. (Before the matter could be settled in Congress or a courtroom, the effort imploded when a security researcher at Bell Labs found a fatal flaw in the Clipper scheme.[9])

The RSA T-shirt was the most visible example of mid-1990s activist art intended to illustrate the futility of trying to control software. Of course, the design files were posted online so that anyone could make their own. A few particularly dedicated cypherpunks even had themselves tattooed with the RSA Perl code, putting a new technological spin on the old line about making your body into a deadly weapon.

On the one hand, turning RSA's elegant math into a digital Jackson Pollock slightly undercut the freedom-of-speech argument that software is a medium of expression for programmers to communicate with each other. The Perl implementation is notoriously, almost incomprehensibly compact. On the other hand, compressing the algorithm so it fit neatly on a shirt helped make the central point. Software, unlike a surface-to-air missile, can fit anywhere. Trying to control it is like trying to control T-shirts. (Plus, it left room on the back for text from the Bill of Rights with a large red VOID stamp, in case the political point wasn't obvious already.)

In 1996, a federal court ruled for Bernstein, and its decision was upheld on appeal in 1999: Snuffle's source code was protected under the First Amendment. The government asked the court to reconsider the case and, while it was pending, substantially weakened the export-control rules for cryptography, mooting his case. The rules aren't entirely gone, though: software licenses still tell you not to send your copy of macOS to embargoed countries (such as Iran and North Korea), and software companies still pay lawyers to fill out export license application forms.[10]

The software-is-speech argument has outlasted the particular debates that produced the RSA T-shirt. In the early 2000s, the movie industry attempted to suppress DeCSS, a program to decrypt DVDs. In response, activists turned its code into an animated *Star Wars* crawl, a square dance, an epic poem composed of 456 haiku, and yes, a T-shirt. They were gathered together in a “Gallery of CSS Descramblers” used to make the case that software could be expressive speech.[11] They made their point but lost the case; courts held that yes, software could have expressive elements, but that speech interest could be steamrolled in the name of copyright.[12] More recently, similar issues have come up in attempts to restrict files for 3D-printed firearms.[13]

Although it is too soon to declare that the crypto wars are over and crypto won, today strong encryption is everywhere. It’s built into every major web browser and every major messaging app. Banks use it, activists use it, as do cops and criminals, software developers and users, and teenagers sending each other Minecraft memes.

How about Adam Back, who created the RSA T-shirt? He runs the blockchain company Blockstream, which has raised $100 million to build technologies resistant to government censorship.[14] That’s a lot of T-shirts.

17

Needles in the World's Biggest Haystack

The Algorithm That Ranked the Internet

John MacCormick

We've all experienced, from time to time, a kind of compulsive web surfing in which we follow link after link, browsing content that becomes less and less relevant to the task at hand. This happened to me only yesterday: while working on some artificial intelligence research, I clicked on something interesting, followed a few links, and 20 minutes later found I was deep into an article about the human brain and consciousness. Strangely enough, this "random surfer" model of Internet browsing also lies at the heart of one of the most revolutionary pieces of code to impact the Internet age: Google's PageRank algorithm.

It is widely believed that the PageRank algorithm, invented and first published by Google cofounders Sergey Brin and Larry Page in 1998, was the single most important element in launching the Google search engine to its dominance of the emerging web search industry in the early 2000s. Around this time, Google leapfrogged some established players such as Lycos and AltaVista, which have since faded into obscurity. How and why did this happen? The key insight of the Google cofounders was that a web search engine would live or die according to the quality of its *ranking* of search hits. The technology of crawling and indexing the entire web was already well understood—Lycos, AltaVista, and others had mastered that. The problem was that

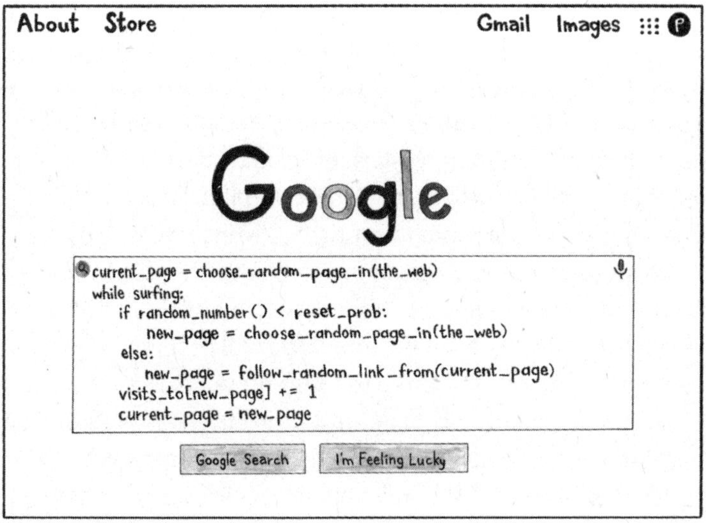
About
Store
Gmail
Images
Google
current_page = choose_random_page_in(the_web)
while surfing:
if random_number() < reset_prob:
new_page = choose_random_page_in(the_web)
else:
new_page = follow_random_link_from(current_page)
visits_to[new_page] += 1
current_page = new_page
Google Search
I'm Feeling Lucky

Code that simulates Google's PageRank algorithm.

most search queries would overwhelm the user with far too many hits. For example, if I search the Web these days for "field hockey," there are more than 300 million hits. This is only a tiny fraction of the entire Web, but still far too large to be a useful set of results. A good search engine, therefore, needs to *rank* those 300 million pages. Ideally, they would be ranked so that the top three to five search results are highly authoritative and informative about field hockey. With their 1998 PageRank algorithm, Brin and Page thought they had figured out a way to find the most authoritative and informative pages automatically—and the public voted with their mouse clicks. Google's results were far more relevant than those of competitors such as Lycos and AltaVista. Google's market share soared, and a twenty-first-century Internet giant was born.

The concept of a web page being "authoritative" is subjective. It's a property that knowledgeable humans can assess, but could a computer algorithm exhibit this same level of understanding and judgment? At first glance, this is a problem that requires artificial intelligence. It's tempting to attack the problem with the latest neural network or machine learning techniques, because these tools can learn how to classify inputs based on examples. The insight of Brin and Page was that we don't really need AI here. Instead, we can harvest—some would say *crowdsource*—the opinions of billions of humans all over the planet to automatically infer their opinions about the authoritativeness of web pages.

Suppose you are the author of a web page about field hockey. Over time, you will develop expertise about which other websites are authoritative on the topic. You will probably select the best ones and provide links to them on your own web page. Thousands of other web-page authors around the globe will do the same thing. On average, the best field hockey pages will be the ones that receive the most incoming links from other web pages. This is the first algorithmic principle behind PageRank: pages with many incoming links are likely to be authoritative. Therefore, pages with more incoming links should be ranked higher than pages with fewer incoming links.

But we can do even better. We know that some authors have better judgment than others. For example, the author of a really popular sports web page is likely to be a good judge of whether other sports web pages are authoritative. On average, we should trust this author's opinion more than that of an amateur sports blogger whose page has limited popularity. This is the second algorithmic principle behind PageRank: incoming links from authoritative pages should count for more than incoming links from pages with low authority. To achieve this, we add up the authority scores of a page's incoming links and use this total score to rank hits that match a query. It is this total authority score of the incoming links that Brin and Page defined as the PageRank score of a given page.

You may have noticed an unfortunate chicken-and-egg problem here. The PageRank score of a page *P* depends on the PageRank scores of pages linking to *P*. But the scores of *those* pages depend on their own incoming links, and so on. This produces an infinite chain of computations. Brin and Page knew a mathematical trick that transforms this infinite chain into a straightforward calculation. When we transform the above PageRank definition into mathematics, the scores of a collection of web pages satisfy a certain equation expressed in matrix algebra. That equation takes a form that is familiar to any undergraduate math major: it is a so-called *eigenvalue problem*, and it can be solved efficiently on modern computers even for matrices that encode the links among billions of web pages. So Brin and Page were able to code up the eigenvalue problem, solve it, and use the resulting PageRank values to rank the query hits in their search engine prototype. The rest is history.

But where does the random surfing come in? I opened this essay with the claim that the indulgent time-wasting of a person randomly surfing the Web mimics the algorithm behind Google's rise to power. Well, if your surfing is truly random, you are computing Brin and Page's chicken-and-egg PageRank values! For the math to work correctly, we need to assume you start on a page selected at random from the entire World Wide Web. You randomly choose one of the links on this page and follow it. That leads to a new page where the process is repeated: a random link is selected and followed. This process is repeated many times, with one variation. Every so often, you get bored and perform a *reset* by restarting your surfing at a new random page selected from the whole Web. In their original 1998 publication, titled "The Anatomy of a Large-Scale Hypertextual Web Search Engine," Brin and Page suggested using a reset probability of 15 percent.[1] This means that 85 percent of the time, you follow a randomly selected link, and 15 percent of the time you start at a new random page from anywhere on the Web.

The code in the figure on page 109, written in the programming language Python, shows how the random surfing algorithm can be implemented in a computer program. After a few billion iterations of the surfing strategy, the number of times a given page was visited is a good statistical estimate of that page's PageRank score.

In the decades since PageRank was first deployed, there have been debates as to how important or revolutionary it really is. Was PageRank good math, or just good marketing? Specialists in information retrieval were aware of the basic technique many years earlier, and a search engine called RankDex was already using a similar style of link analysis to rank pages when Google got started. In addition, search engine companies quickly discovered that PageRank alone was not enough to stay ahead of the competition. Indeed, PageRank provides only a *static* score for each page, meaning that the score depends only on the page itself and not on the query entered by the user. Modern search engines employ *dynamic* scores that are query-dependent, and PageRank is only one component of such scoring systems.

Despite these counterarguments, there are compelling reasons to believe the hype and mystique surrounding PageRank. Brin and Page were among the first to apply the concept at Web scale, where its impact turned out to be immense. And it does seem likely that PageRank was an important contributor to the quality of Google's search results in the early years, when Google's competitors were playing catch-up and desperately trying to improve their own search quality. Ultimately, however, these lines of code—the PageRank code—achieved much more than the launch of a single company from a garage in Menlo Park, California, to its status as a global internet superpower less than a decade later. The PageRank code transformed twenty-first-century society, creating a vast new ecosystem based on web search, enabling us to find relevant needles in gigantic haystacks, and revolutionizing the way we interact with information.

18

A Failure to Interoperate

The Lost Mars Climate Orbiter

Charles Duan

The Mars Climate Orbiter was falling fast. After nine months of flight, the spacecraft had reached the red planet on its target date of September 23, 1999, and was preparing to enter Mars' orbit in order to collect weather data from the planet's atmosphere. But while the orbiter was supposed to fly at least 160 kilometers above the Martian surface, the latest estimates suggested it was at 110 kilometers and dropping. At less than 98 kilometers, the spacecraft would be blasted with damaging heat; at less than 85, it would be rendered unsteerable. The navigation crew at NASA's Jet Propulsion Laboratory (JPL) began firing the orbiter's thrusters to keep it in position, but to no avail. At 9:05 UTC, it entered Mars' shadow, which blocked radio communication with the ground crew. It was never heard from again, likely having burned upon entry into Mars' atmosphere. The final estimated altitude: 57 kilometers.[1]

In the weeks that followed, NASA and its coordinate engineering teams at JPL and Lockheed Martin Aeronautics scrambled to explain the loss of the $145 million spacecraft. They and other investigators produced at least five post-mortem reports on the error, variously pointing fingers at management, budgets, and team communication. The one point of consensus among all of them was that the disaster would have been averted but for a software bug: a missing line of code that should have multiplied a number by 4.45.[2] That missing line of code

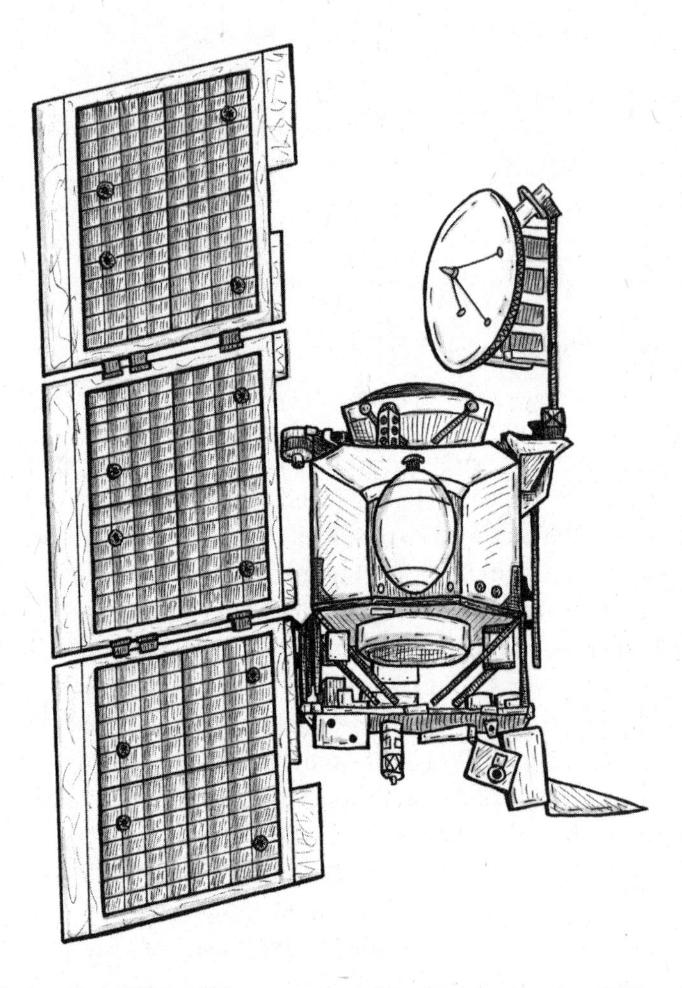

The lost Mars Climate Orbiter, prior to crashing. (Drawing based on 1999 press materials from NASA.)

and its destruction of a multimillion-dollar spacecraft would foreshadow a systemic problem with the burgeoning software industry—a problem that would persist for more than 20 years.

The missing code was supposed to help manage the Mars Climate Orbiter's orientation during flight. About twice a day, the small spacecraft would execute an operation called "angular momentum desaturation," in which it fired small thrusters to orient the spacecraft in the right direction. Unfortunately, the thrusters had the side effect of pushing the spacecraft off-course, changing its velocity by a tiny amount—less than a ten thousandth of a percent.[3]

Though these delta-V's, as the scientists called them, were minuscule, they were also significant because they fed into the larger computation of determining where the orbiter was at any given time. Navigating a spacecraft is hard—there are no street signs in space, and measurements based on star positions or Earth-based radar are incomplete and inaccurate at best. Instead, JPL kept track of all the forces acting upon it—the pressure of solar wind and the gravity of the sun and planets, in addition to the delta-V's—and predicted the orbiter's trajectory by adding up the forces.[4] This perhaps sounds like predicting how far a car has driven by tallying up presses on the gas and brake. But just two years earlier, JPL had used largely the same navigation modeling software to successfully place the Mars Global Surveyor into orbit. The two spacecraft differed significantly in shape, intended motion, and thruster configuration, meaning that the navigation modeling software for the Surveyor would not work as-is for the Mars Climate Orbiter.[5] But the physics was the same, and so the Mars Climate Orbiter's engineers expected that they could reuse the Surveyor's modeling software, just feeding it slightly modified input data tailored to the Orbiter.

Yet it was what fed into that modeling software that ultimately sank the Mars Climate Orbiter. Upon each of the twice-daily angular momentum desaturation events, a software program called `SM_FORCES` would run. That program, written by JPL's contractor Lockheed Martin, would produce files of thruster data for JPL's modeling software to use in its computations.[6] But while JPL's software expected the data files to use metric units of Newton-seconds, `SM_FORCES` produced data files in imperial units of pound-force-seconds.[7] As a result, the numbers were too small by a factor of 4.45, so JPL predicted the orbiter's delta-V's to be less than reality. While each individual error was small, they added up: After the crash, JPL estimated that the total error over the mission would have been over 10,000 kilometers.[8] Even after several course corrections in flight, 169 kilometers of error remained in the days before the orbiter reached Mars. That distance was tiny compared with the 669 million kilometers the orbiter had traveled,[9] but given the small window for a successful Mars orbit, it was enough to turn the mission from success to failure.[10]

Failure to multiply by 4.45 was no doubt a software bug, but which line of code was buggy—and whose responsibility was it? Lockheed Martin's software could have performed the conversion, but so could JPL's. Each side laid blame on the other: NASA and JPL pointed to a software requirements document that called for metric units,[11] while Lockheed Martin's engineers countered that they had written their software to match sample hand-computed data that JPL had provided—data that apparently used imperial.[12] And in a sense, both programs were correct: `SM_FORCES` produced numbers exactly in line with what its coders intended, and JPL's software ingested and processed data exactly the way its coders had specified. The bug could be said to lie not with either piece of code, but somewhere in between.

That in-between space is the world of *interfaces*—the places *inter* two *faces* of software. Across that space, one program can send information or commands to another. But those transmissions are meaningless 1s and 0s unless the programs have a shared agreement on what they mean, just as two humans must have a shared understanding of a language before they can talk with each other. And even more so than human languages, the "languages" with which computer programs communicate must be precisely defined by agreement between the programs' coders, including the choice of metric versus imperial units. Programs that do so are said to be compatible, or *interoperable*; those that fail will produce wrong results, or, perhaps, crash, literally or figuratively.

Interoperability is what allows the massive tangle of interfaces that we call the Internet to function—to retrieve Emails, web pages, and "All Your Base" memes. As Harvard scholars John Palfrey and Urs Gasser have argued, interoperability born from that early Internet age has enabled consumer choice, encouraged technological progress, promoted competitive markets, and opened up efficiencies and conveniences that have made the highly connected society in which we live today.[13]

Today, interoperability is practically the expectation. Indeed, much frustration with technology is due to failures of interoperability—when Microsoft Word can't open an Apple Pages file, for example.[14] (Apparently a file format incompatibility also plagued the Mars Climate Orbiter mission, forcing one poor JPL analyst to recompute all of Lockheed Martin's trajectory predictions by hand for the first four months of the mission.)[15]

While the failure of interoperability for the Mars Climate Orbiter was an honest mistake, non-interoperability in technology today often arises more deviously: forced incompatibility designed to secure market domination. The great virtue of interoperability is substitutability. JPL could switch to any other contractor's compatible delta-V files with the click of a keyboard. But what if Lockheed Martin could "own" that compatible format, forcing competitors to generate only incompatible files—perhaps using even more exotic units of impulse, such as slug-feet per second or poundal-hours? One can only imagine how much more confusing and disaster-prone a Mars spacecraft operation could become in such a world where JPL must rewrite its software for every new contractor.

Even more worryingly, companies ranging from electronic health care to social media have begun to eschew the late-1990s embrace of interoperability, preferring instead the walled gardens of curated app stores and vertically integrated products designed to lock in consumers and markets.[16] Where once the norm was highly compatible Internet technology, now even a coffee machine may refuse to interoperate with third-party coffee pods.[17]

This shift away from interoperability is troubling. As with space exploration, the digital economy depends on a diverse array of contributors: large firms to be sure, but also startups, small creators, forum frequenters, scientists, engineers, and many more. Interoperability is the glue that enables this diverse community to work together innovatively; failures of interoperability, whether accidental or intentional, leave that community frustrated and divided. Should society fail to pay heed to the importance of interoperability, its technological prospects may very well burn up on entry many times over.

19

The Code That Launched a Million Cat Videos

Lowen Liu

According to Colin Angle, the CEO and cofounder of iRobot, the Roomba faced some early difficulties before it was rescued by two events. The disc-shaped robot vacuum had gotten off to a hot start in late 2002, with good press and a sales partner in the novelty chain store Brookstone. Then sales started to slow, just as the company had spent heavily to stock up on inventory. The company found itself on the other side of Black Friday in 2003 with thousands upon thousands of Roombas sitting unsold in warehouses.[1]

Then around this time, Pepsi aired a commercial starring comedian Dave Chappelle.[2] In the ad, Chappelle teases a circular robot vacuum with his soft drink while waiting for a date. The vacuum ends up eating the comedian's pants—*schlupp*. Angle remembers that at a team meeting soon after, the head of e-commerce said something like: "Hey, why did sales triple yesterday?" The second transformative moment for the company was the rapid proliferation of cat videos on a new video-sharing platform that launched at the end of 2005. A very specific kind of cat video: felines pawing suspiciously at Roombas, leaping nervously out of Roombas' paths, and, of course, riding on them. So many cats, riding on so many Roombas. It was the best kind of advertising a company could ask for: it not only popularized the company's product but made it charming. The Roomba was a bona fide hit.

By the end of 2020, iRobot had sold 35 million vacuums,[3] leading the charge in a booming robot vacuum market.

The Pepsi ad and the cat videos appear to be tales of early-days serendipity, lessons on the power of good luck and free advertising. They also appear at first to be hardware stories—stories of cool new objects entering the consumer culture. But the role of the Roomba's software can't be underestimated. It's the programming that elevates the round little suckers from being mere appliances to something more. Those pioneering vacuums not only moved, they *decided* in some mysterious way where to go. In the Pepsi commercial, the vacuum is given just enough personality to become a date-sabotaging sidekick. In the cat videos the Roomba isn't just a pet conveyer, but a diligent worker, fulfilling its duties even while carrying a capricious passenger on its back. For the first truly successful household robot, the Roomba couldn't just do its job well; it had to win over customers who had never seen anything like it.

Like many inventions, the Roomba was bred of good fortune but also a kind of inevitability. It was the brainchild of iRobot's first hire, former MIT roboticist Joe Jones, who began trying to make an autonomous vacuum in the late 1980s. He joined iRobot in 1992,[4] and over the next decade, as it worked on other projects, the company developed crucial expertise in areas of robotics that had nothing to do with suction: it developed a small, efficient multithreaded operating system; it learned to miniaturize mechanics while building toys for Hasbro; it garnered cleaning know-how while building large floor sweepers for SC Johnson; it honed a spiral-based navigation system while creating mine-hunting robots for the US government.[5] It was a little like learning to paint a fence and wax a car and only later realizing you've become a Karate Kid.

The first Roombas needed to be cheap—both to make and (relatively) to sell—to have any chance of success reaching a large number of American households. There was a seemingly endless list of constraints: a vacuum that required hardly any battery power, and navigation that couldn't afford to use fancy lasers—only a single camera.[6] The machine wasn't going to have the ability to know where it was in a room or remember where it had been. Its methods had to be *heuristic*, a set of behaviors that combined trial and error with canned responses to various inputs. If the Roomba were "alive," as the Pepsi commercial playfully suggested, then its existence would more accurately have been interpreted as a progression of instants—*did I just run into something? Am I coming up to a ledge? And if so, what should I do next?* All conditions prepared for in its programming. An insect, essentially, reacting rather than planning.

And all this knowledge, limited as it was, had to be stuffed inside a tiny chip within a small plastic frame that *also* had to be able to suck up dirt. Vacuums, even handheld versions, were historically bulky and clumsy things, commensurate with the violence and noise of what they were designed to do. The first Roomba had to eschew a lot of the more complicated machinery, relying instead on suction that accelerated through a narrow opening created by two rubber strips, like a reverse whistle.

But the lasting magic of those early Roombas remains the way they moved. Jones has said that the navigation of the original Roomba appears random but isn't—every so often the robot should follow a wall rather than bounce away from it.[7] In the words of the original patent filed by Jones and Roomba cocreator Mark Chiappetta, the system combines a *deterministic component* with *random motion*.[8] That small bit of unpredictability was pretty good at covering the floor—and also made the thing mesmerizing to watch. As prototypes were developed, the code had to account for an increasing number of situations as the company uncovered new ways for the robot to get stuck, or new edge cases where the robot encountered two obstacles at once.[9] All that added up until, just before launch, the robot's software no longer fit on its allotted memory. Angle called up his cofounder, Rodney Brooks, who was about to board a transpacific flight. Brooks spent the flight rewriting the code compiler, packing the Roomba's software into 30 percent less space.[10] The Roomba was born.

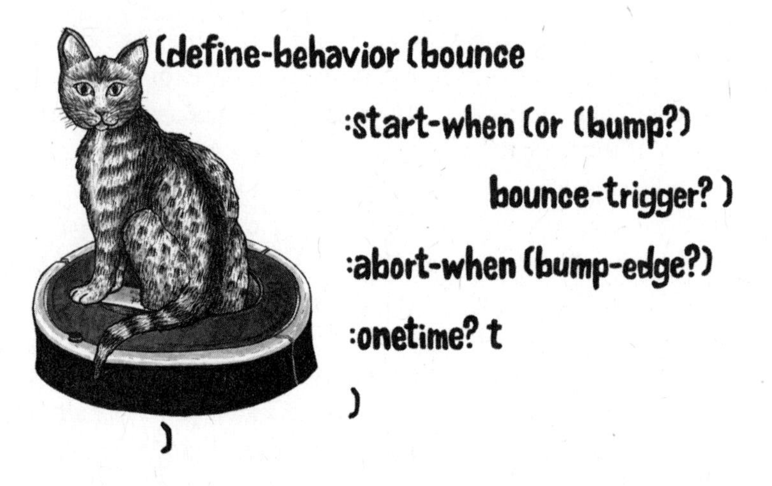

A snippet of code at the heart of the Roomba's navigation system. The robot scoots along the floor, constantly checking for inputs—*Have I bumped into something? How about now? And now?*—and reacting in ways that are both predetermined and a little mysterious.

In 2006 Joe Jones moved on from iRobot, and in 2015 he founded a company that makes robots to weed your garden.[11] The weeding robots have not, as yet, taken the gardening world by storm. And this brings us to perhaps the most interesting part of the Roomba's legacy: how lonely it is.

You'd be in good company if you once assumed that the arrival of the Roomba would open the door to an explosion of home robotics. Angle told me that if someone went back in time and let him know that iRobot would build a successful vacuum, he would have replied, "That's nice, but what else did we really accomplish?" A simple glance around the home is evidence enough that a future filled with robots around the home has so far failed to come true. Why? Well for one, robotics, as any roboticist will tell you, is hard. The Roomba benefited from a set of very limited variables: a flat floor, a known range of obstacles, dirt that is more or less the same everywhere you go. And even that required dozens of programmed behaviors.

As Angle describes it, what makes the Roomba's success so hard to replicate is how well it satisfied the three biggest criteria for adoption: it performed a task that was unpleasant; it performed a task that had to be done relatively frequently; and it was affordable. Cleaning toilets is a pain but not done super frequently. Folding laundry is both, but mechanically arduous. Vacuuming a floor, though—well, now you're talking.

Yet for all the forces that led to the creation of the Roomba, its invention alone wasn't a guarantee of success. What *is* it that made those cat videos so much fun? It's a question that lies close to the heart of the Roomba's original navigation system: part determinism, part randomness. My theory is that it wasn't just the Roomba's navigation that endeared it to fans—it was how halting and unpredictable that movement could be. The cats weren't just along for an uneventful ride; they had to catch themselves as the robot turned unexpectedly or hit an object. (One YouTuber affectionately described the vacuum as "a drunk coming home from the bar.") According to this theory, it's the imperfection that is anthropomorphic. We are still more likely to welcome into our homes robots that are better at slapstick than superhuman feats. It's worth noting that the top-of-the-line Roomba today will map your rooms and store that map on an app, so that it can choose the most efficient lawnmower-like cleaning path. In these high-end models, the old spiral navigation system is no longer needed. Neither is bumping into walls.

Watching one of these Roombas clean a room is a lot less fun than it used to be. And it makes me wonder what the fate of the Roomba may have been had the first ever robot vacuum launched after the age of smartphones, already armed with the capacity to roll through rooms with precise confidence, rather than stumble along. It's not always easy, after all, to trust someone who seems to know exactly where they are going.

20

Nakamoto's Prophesy

Bitcoin and the Revolution in Trust

Quinn DuPont

When the person known as Satoshi Nakamoto started working on Bitcoin in 2007, then-US Federal Reserve Chair Alan Greenspan was already warning of a looming recession, and banks in England were facing a serious liquidity crisis. For years, banks had been bingeing on cheap credit and were selling mortgages to unqualified borrowers. In late 2008, Nakamoto published the now-famous "Bitcoin: A Peer-to-Peer Electronic Cash System" white paper that laid out the design for a novel cryptocurrency.[1] By then, the global economic crisis was in full swing. When the first bitcoins were "minted" in January 2009, Nakamoto made the system's political mission crystal clear: indelibly encoded in the so-called Genesis block of transactions, Nakamoto typed out a London newspaper's headline, "The Times 03/Jan/2009 Chancellor on brink of second bailout for banks." With a stroke, Nakamoto had timestamped Bitcoin's origin and marked its target.

Nakamoto was keenly aware of the modern nation-state's monopoly on the creation of money, which is maintained by taxation laws, anticounterfeit laws, and banking regulations, but Nakamoto also knew that most money is de facto created by commercial banks through their lending practices. By developing a peer-to-peer digital money beyond the remit of nation-states and their banks, Nakamoto was championing the legacy of cypherpunks—information-age anarchists who sought their own micro-nations and digital cash. But Nakamoto also sensed

Satoshi Nakamoto included a hash of the *Times* headline from January 3, 2009 in the Genesis block of Bitcoin.

a broader, more socially transformative use. Bitcoin was to spur a revolution in trust.

The original Bitcoin whitepaper was focused on Internet commerce. Nakamoto wrote:

> Commerce on the Internet has come to rely almost exclusively on financial institutions serving as trusted third parties to process electronic payments. While the system works well enough for most transactions, it still suffers from the inherent weaknesses of the trust based model. Completely non-reversible transactions are not really possible, since financial institutions cannot avoid mediating disputes. The cost of mediation increases transaction costs, limiting the minimum practical transaction size and cutting off the possibility for small casual transactions, and there is a broader cost in the loss of ability to make non-reversible payments for non- reversible services.[2]

Nakamato believed that without irreversible transactions, financial services would need to collect ever more information to be able to trust their customers, in turn reducing privacy. With Bitcoin, however, trust would be replaced by cryptographic proof, which would protect sellers from fraud and buyers from unscrupulous sellers through secure, automated escrow and smart contracts. Since Bitcoin transactions are (pseudonymously) recorded in a shared ledger, anyone can inspect the ledger to confirm payment. For the seller, the shared ledger offered protection, since transactions cannot be changed or removed from the ledger; payment is confirmed before shipping. For the buyer, the shared ledger can serve as a third party for escrow, finalizing payment only once the shipment is complete.

As the global economy recovered, Nakamoto and a small team of open-source software developers continued to work on Bitcoin. At first, Bitcoin was a fun hobby for most, and each coin was nearly worthless. (Famously, two pizzas sold for 10,000 bitcoins on May 22, 2010, now celebrated as "Pizza Day.") Then, in 2011, with no warning and little fanfare, Nakamoto disappeared. Over the years, open-source software developers continued to enhance Bitcoin while the news media hunted for Nakamoto's whereabouts. Remarkably, none of the estimated one million bitcoins held in Nakamoto's wallet have been spent (now worth at least USD $50 billion).

Bitcoin has had a fascinating and tumultuous history that spawned copycats, a blockchain "revolution," and even—with unacknowledged irony—Central Bank digital currencies. The success of Bitcoin can be attributed to many factors, but in the Bitcoin whitepaper Nakamoto singled out the "inherent weaknesses of the trust based model" for electronic commerce. Nakamoto felt that trusted third parties like payment processors were costly and inefficient and, worse still, could not actually be trusted to act impartially. Nakamoto's suspicions were validated in 2011 when, buckling to US political pressure, Visa, Mastercard, and other financial organizations froze the accounts of the whistleblowing website Wikileaks. By supplanting trust with cryptographic proof, Bitcoin made it possible to securely interact without a mediating third party.

Long before ubiquitous computing and global telecommunications, the sociologist Georg Simmel observed that trust—faith in others to not deceive—was crucial to modern life.[3] Foreshadowing the emergence of Bitcoin, Simmel observed how trust relationships are usually reciprocal and, increasingly, accounted for by money. Indeed, more than a hundred years ago he observed how money produced a social "credit-economy" where obligations are measured by money and supported by trust. Another early sociologist, Max Weber, noticed that the fledgling capitalist economy had progressed by instituting a more reliable calculation of obligations through improvements in bookkeeping, business practices, and technology.[4] With Bitcoin, Nakamoto seemed to be suggesting, perfect accounting could eliminate the need for trust altogether.

Around 2014—after Nakamoto had vanished—software developers started to realize that Bitcoin's underlying ledger system could be extended to situations beyond money. This ledger system, called blockchain, meant that even in low-trust environments, complex and robust kinds of interaction could be achieved. For instance, decentralized finance (DeFi) now allows speculators to trade assets directly, without the commission and regulatory overhead of centralized exchanges. Today, billions of dollars in assets—including artwork, property, and cryptocurrencies—are traded this way.

In Bitcoin, the underlying blockchain is necessary to create agreement among globally distributed ledgers. Since Bitcoin has no central server (a "third party") to order and process transactions, so-called miners compete in a lottery to validate and bundle transactions into blocks. Using a "gossip" protocol, each miner listens for incoming transactions (sent from individual wallets) and computes an algorithm that produces a random result. If the miner discovers a random "winning" result (a parameter set by the protocol), it seals the transactions into a block, chains it to the previous one, and is rewarded for the trouble with newly minted bitcoins. Because the bitcoin reward is valuable, miners are incentivized to run the protocol and act honestly.

So, if a *dishonest* person wanted to submit a fraudulent transaction to the Bitcoin network, a colluding miner would still be required to compute the expensive, power-consuming algorithm. Therefore, counterfeiting is practically impossible on the open Bitcoin ledger because the honest majority of miners would reject the transaction and its mining reward, making collusion economically irrational. But since there is no centralized authority to determine valid from fraudulent blocks, how does the network weed out fraudsters who only attempt to alter existing transactions, or "double spend" coins? When presented with competing ledgers and two equally plausible versions of truth, fraudulent blocks are detected and rejected by honest miners because they always select the longest chain—the one with the most "work" behind it. Since existing transactions are from the past and therefore shorter in length (they have less work), the majority of honest miners will come to agreement and reject the fraudulent chain.

This remarkable process of consensus keeps Bitcoin payments secure when dealing with adversaries. Even though digital code can be endlessly copied and pasted, the blockchain prevents counterfeiting and double-spending. Security is assured because the Bitcoin network financially encourages miners to check and include all valid transactions and eliminates payment reversals. Since the lottery chooses a winner at random, censors and cheaters gamble in a loser's game. The blockchain achieves consensus and agreement about a single version of truth through a sophisticated dance of cryptography and economic incentives.

When monetary transactions are replaced with executable code, as in the case of Ethereum and other blockchain platforms, it becomes possible to achieve consensus about any computational result—a decentralized "world" computer. Today, this process is used to secure shipping manifests, for example, as they pass between companies. When combined with environmental sensors and wireless networking (such as industrial-scale Internet-of-Things), a blockchain can detect issues (like an opened container or a failing refrigeration unit) and automatically take action. Even with untrustworthy shippers, when the digital manifest and its sensor data are entered onto the blockchain, the records cannot be altered or erased. For these kinds of applications, blockchain abstracts away the concept of digital money and extends the Bitcoin whitepaper's original use case to much richer business environments.

Consensus about the result of executed code can also support multistakeholder, low-trust environments to enable better governance. Indeed, achieving consensus about records bolsters trust in ways that may be able to solve collective action problems, including pressing social issues like climate change, corruption, and inequality. For example, since blockchains can securely record the results of voting and automatically take action with smart contracts, a quasi-legal smart contract might, for example, automatically allocate funds for condo repair, where otherwise individual decision-making might result in a tragedy of the commons. These governance mechanisms are the practical realization of decades of development in game theory, which can produce socially optimal results for condo boards, access to water rights, land titling, and many other public or "toll" goods. In recent years, a refinement of game theory called cryptoeconomics has emerged to examine how sophisticated forms of consensus support trust to enable better governance. This fledgling discipline draws on complex systems, mechanism design (known as "reverse" game theory), and information security and may underpin novel, algorithmically enhanced governance platforms.

Today, Nakamoto's original political vision is no longer recognizable; after all, Bitcoin is now largely a Silicon Valley–style casino. But Bitcoin has helped create new values for democracy, financial inclusion, and ownership. The Millennial and Gen Z generations in particular, those who were left behind in the wake of the 2008 global economic crisis that birthed Bitcoin, have eagerly embraced this decade's *Sturm und Drang* by remaking old institutions in their own image. To the consternation of state regulators and politicians, the youth have flocked to Bitcoin and in turn brought new relevance to Nakamoto's prophecy.

21

The Curse of the Awesome Button

Will Oremus

It was summer 2007. Facebook was three years old and growing at a heady pace. Originally for college students, it had opened to the public the previous fall. Now it had 30 million users.[1] What it didn't have was a simple way for them to show interest in each other's posts. The only way to acknowledge a post was to comment on it.

Leah Pearlman, one of Facebook's three product managers at the time, found that inefficient. Popular posts would receive long strings of comments, many just one or two words (such as "awesome" or "congrats"), with no way to locate the interesting ones in a sea of noise. That might not seem like a big problem, but it chafed Pearlman and her coworkers, almost all of whom were in their early twenties and active Facebook users themselves. To Pearlman, the comment chains felt like something you would see on Facebook's more established rival, MySpace, with its anything-goes customizability. That clashed with Facebook's clean design and straightforward functionality.

She and a few other Facebookers—engineering managers Akhil Wable and Andrew Bosworth, designer Justin Rosenstein, and internal communications manager Ezra Callahan—set out to build a universal, seamless way to express approval on the social network. They code-named the project "Props."

What they ended up developing, with help from a succession of other designers and engineers, was a button that became an iconic symbol of Facebook, reshaped the Internet, reconfigured the media, and altered the course of world politics. If they'd

known the "like" button would do all that, its creators say now, they might have thought about it rather differently than they did at the time. The same virtues that made it such an elegant user-interface solution for a fast-growing software startup in 2007—its simplicity, its ease of use, its universality, its convenience as a metric of value—eventually came to look like vices when applied on a grand, societal scale by a globally dominant online platform.

"I felt like I was cleaning the garage by adding the like button," Pearlman told me.[2] "I never expected what happened.... We were working on other things at the time that felt like a bigger deal by far."

One obvious question is whether the Facebook like button's creators should have anticipated those long-term consequences. A more interesting, haunting question lurks behind that one: Would they, or could they, have done anything differently even if they'd known?

That Facebook would eventually give users an easy way to express affirmation was perhaps overdetermined. By 2007, a handful of smaller platforms had already launched approval buttons of various sorts. And Pearlman wasn't the only one at Facebook thinking along those lines. Rosenstein told *The Ringer* in 2017 that he had been looking for "a way to increase positivity in the system" to help Facebook create "a world in which people uplift each other rather than tear each other down."[3]

But just what shape Facebook's version of "props" should take, and how it should work, proved contentious. Pearlman added the "awesome button," as the group initially called it, to Facebook's internal ideas board, and it got enough votes from their coworkers to spur a "hackathon"—an all-night coding session in which engineers and designers cobble together prototypes of potential new features. On July 17, 2007, a team consisting of Bosworth, Rebekah Cox, Ola Okelola, Rosenstein, and Tom Whitnah coded the first awesome button, according to a detailed account that Bosworth posted to the question-and-answer site Quora in 2010.[4] It was well-received and got the green light for development.

As Bosworth recalled it, the button generated excitement across the company.[5] The ads team thought it could be used to show people better ads. The platform team thought it could be used to filter out bad apps. The news feed team, which Bosworth led, reckoned it could help them decide which posts to show in people's feeds. (At that point, the main factor in feed ranking was the number of comments a post received.) The button's seemingly universal applicability meant that it had to be versatile and appropriate across all sorts of contexts. And so seemingly trivial design decisions took on crushing weight. Was "awesome" the right name? Should there be a corresponding minus sign or thumbs down? Should it come with a counter?

Designing a button that would be both visually simple and instantly understandable was "a really interesting problem," Pearlman said.[6] "Because remember, there weren't really buttons on the Internet" back then. (Digg's thumbs-up icon and Vimeo's like button were among the few precursors;[7] the smaller social network Friendfeed launched a like button in 2007,[8] though the Facebookers who were working on the awesome button at the time insist they hadn't noticed it.)

After delays, the team presented the awesome button to CEO Mark Zuckerberg for final approval in November 2007. Zuckerberg surprised them by rejecting it. In Bosworth's version of the story, Zuckerberg saw potential conflicts with Facebook's privacy defaults (would likes be public or private?), a forthcoming ad platform (the ill-conceived and ill-fated Facebook Beacon), and the share button that Wable's team was working on. He also dissented on the name, preferring "like" to "awesome."[9]

Callahan remembers at least a couple people inside Facebook raising a concern that would prove prescient: they worried, he told me, that "low-effort directional feedback," in the form of a "like" or "awesome" button, would "eliminate thoughtful engagement, because people were lazy and would take the lazy way out" if given the option.[10]

That might sound like a very post-2017 critique: the notion that mindless clicks ought to be eschewed in favor of meaningful interactions for the sake of democratic discourse and our own well-being. But it's not clear that anyone at Facebook back then was laboring under the assumption that their product design decisions would have world-historical implications. Rather, they were focused squarely on building a better and more engaging product than MySpace, and the crucial question was whether replacing comments with clicks would aid or inhibit that quest. The implicit assumption, according to Pearlman, was that what was good for Facebook's users was good for Facebook's business, and probably good for the world.[11]

That set of assumptions—that tech startups were underdogs, that they were forces for good, and that their success in business would naturally coincide with bettering the world—was common in Silicon Valley at the time. Think of Google's erstwhile unofficial motto: "Don't be evil." But that blithe moral confidence bred moral complacency. When you assume your good intentions will entail good outcomes, provided only that you succeed in outmaneuvering and outhustling your rivals, then stopping to think too hard about what could go wrong is not only unnecessary, it's potentially ruinous, because it could prevent the good guys—that is, you—from succeeding. (Google removed "Don't be evil" from its code of conduct in 2018.[12])

For a time, it looked as though inertia might defeat Facebook's "props" initiative. By late 2008, people started joking about the "curse of the awesome button," said Soleio Cuervo, a former Facebook designer who worked on it.[13] "There was a lot of concern internally that 'liking' was going to cannibalize engagement."

What saved the like button was, in true Silicon Valley fashion, an appeal to data. In a test, Facebook data analysts found that popular posts with the button actually prompted more interactions than those without. Bosworth has suggested that this may have been at least partly because likes on a post became a signal for the news feed algorithm to show that post to more people.[14] That finding turned out to be decisive. By February 2009, Zuckerberg had approved a final version of the like button, drawn as a thumbs-up by Cuervo in Photoshop.

But coding the like button involved much more than just drawing it. Each like had to be stored in databases that linked it to both the post itself and the person doing the liking. Much of the coding was done by engineer Jonathan Pines, with contributions from another engineer, Tom Whitnah, and product manager Jared Morgenstern.[15]

On February 9, 2009, Pearlman announced the like button's launch with a Facebook Note—the company's version of a blog post—titled "I like this."[16] She encouraged users to deploy it liberally, and users complied. The like button was an instant hit, and Facebook soon found ways to ingratiate it into the fabric of not just its platform, but the Internet beyond. By the following year, you could like people's comments as well as their posts.[17] The like button also became the default way to follow publishers and brands on Facebook—and when you did, Facebook would use your like to advertise those same pages to your friends. In April 2010, Facebook unveiled a suite of "social plug-ins"—including the like button—that allowed people to "like" pages outside of Facebook itself.[18] Years later, those plug-ins became ad-tracking beacons, telling Facebook whenever a logged-in user visited a site that sported a like button, so that Facebook could use that information to target ads.[19]

The like button quickly became far bigger than even its biggest-thinking creators had envisioned. Likes became expressions of taste and identity. They became the driving force in an increasingly potent and complex news-feed ranking algorithm: the more likes a post got, the more people Facebook would show it to. They became an asset for brands and advertisers. They became a rich source of data for Facebook itself, telling the company about each user's preferences and browsing habits.

Perhaps most transformatively, likes became a powerful incentive for users of Facebook—a group that grew over the years to nearly three billion.[20] The like counter on each post became an explicit measure of its popularity, and an implicit measure of its value. Consciously or otherwise, users learned what sorts of posts would rack up likes and wrote more like that, while learning to avoid those that garnered only a few. Those users included not just ordinary individuals, but public figures, corporate brands, and media companies. Facebook's news feed algorithm became the single most influential distributor of information in many societies, and it spawned imitators, from Instagram to Twitter to TikTok, until likes were everywhere.

In terms of sheer impact, the like button was one of the most successful pieces of code ever shipped. But when you examine the quality of that impact, its flaws become glaring. In building a feature whose primary function was to simplify the terms of interaction with online content, Facebook understood that it risked cheapening engagement. What it failed to anticipate was how it might lead to a cheapening of the content itself. The posts that thrived uniquely in Facebook's feed, fueled by quick-twitch likes, were the ones that offered simple solutions to complex problems, that offered scapegoats, that played on people's biases and manipulated their emotions. Those that addressed the same problems with nuance, with ambivalence, withered.

If failing to generate likes just meant losing a popularity contest on Facebook, likes might have been tolerable. But the starring role that the company gave likes in the news feed algorithm meant that the popularity contest was not one you could opt out of. If your posts didn't generate likes, they would be hidden from people's feeds—even your own followers' feeds—and replaced by the posts of others who were all too willing to play the manipulation game. Media companies laid off journalists[21] or even went out of business[22] because they couldn't compete on Facebook, or because they refused to change their coverage in the ways needed to do so. Facebook likely played a role[23] in the demise of local newspapers across the country, the rise and fall of manipulative clickbait sites across the Web, and the shifting of the entire media toward pandering, manipulative discourse. And that discourse, in turn, likely played a role in facilitating the rise of populist politicians expert at pandering to people's biases, fears, and base instincts. Their messaging, like Facebook's like button, works best when you don't think too hard.

Facebook and other social networks are now experimenting with removing like counters, or hiding them from public view.[24] Whether that will undo any of the damage is not yet clear. Pearlman, for one, believes the real mistake was the counter; the button on its own, she told me, would have been more benign.

So do she and the like button's other creators regret developing it? Not exactly. It seems that most have come to believe, or at least tell themselves, that it was more or less inevitable. Pearlman believes the harmful effects reflect human nature: "It's my belief we needed to play this out all the way, to see it and decide to choose something else."[25] Callahan, similarly, suggested that if Facebook hadn't popularized the like button, someone else would have.[26] He acknowledged in the same breath that it feels like a weak excuse.

Still, there's probably some truth in it. Facebook wasn't the only company building social tools with the single-minded goal of connecting as many users as possible and keeping them engaged. Let's say some 2007 company was building social networks around the idea that friction was necessary, that nuance was a priority, that supporting democratic discourse was part of their role, and that doing so might require them to eschew features that would help them grow faster or become more addictive. Presumably, it would have quickly been outcompeted by the others in an industry where scale and network effects are critical to funding, ad revenue, and ultimately survival. It probably wouldn't be the dominant social network today. And the engineers who designed its key features all those years ago, never thinking that their products might have world-historical implications, would have been correct.

22

The Bug No One Was Responsible For—Until Everyone Was

Josephine Wolff

On April 7, 2014, Colm MacCárthaigh was at work on the 14th floor of Amazon's Blackfoot building in Seattle when he was paged for an emergency. MacCárthaigh was a principal engineer at Amazon Web Services at the time, and the company had just learned about a new software vulnerability in the code underlying their encryption protocols—code used not just by Amazon but by many other companies and websites worldwide. Dubbed Heartbleed, the bug quickly became headline news as software engineers like MacCárthaigh scrambled to patch their products and services before attackers could take advantage of the vulnerability to steal sensitive information like website credentials or credit card numbers. Amazon organized an emergency response conference call and started rolling out patches as quickly as possible, some within an hour of the call. Five years later, in a Twitter thread, MacCárthaigh wrote that he still recalled the day he learned about Heartbleed "vividly," writing that "[i]t felt like the internet was on fire."[1]

Software vulnerabilities are discovered every day, but Heartbleed was different—it was a bug in a vital piece of security software, used by everyone from major tech firms like Amazon to tiny individual websites. The software affected by the Heartbleed bug was the OpenSSL software library, a repository for code that was used by many websites and other online applications to encrypt traffic sent to and from their users. OpenSSL, which implements the Transport Layer Security (TLS) and Secure Sockets Layer (SSL) protocols to encrypt traffic sent between computers, is widely used by website operators and app developers to encrypt traffic. TLS and SSL are encryption standards that can be implemented using other code besides OpenSSL, but many companies find it helpful to use the OpenSSL implementation rather than write their own from scratch. Part of what makes OpenSSL code useful and popular is that it's not owned by any particular corporate entity—anyone who wants to implement the TLS or SSL protocols can use it without having to purchase a license or get permission. But that also means that there's no company or individual in charge of the code's upkeep—a problem that came to the fore in particularly dramatic fashion in 2014 with Heartbleed.

The Heartbleed vulnerability was discovered by Google researcher Neel Mehta. But its branding—the memorable name and minimalist bleeding-heart logo—came from the Finnish security firm Codenomicon, and it quickly became clear that it was everyone's problem. Because of Heartbleed, devices and services that relied on OpenSSL to encrypt online communications were susceptible to leaking all sorts of sensitive information from passwords and bank account numbers to the contents of Emails. When MacCárthaigh and others first learned of the vulnerability, it was unclear whether anyone had already started exploiting it to steal information. But as soon as the bug became public it was only a matter of time before someone would try. Making matters worse, the bug affected millions of devices, especially Android phones, because OpenSSL was used so widely.[2]

Now, every single company that had used OpenSSL in its products needed to update its code in order to protect its customers. But many did not. And there was no way to force the people who didn't know how to fix the problem, or didn't want to bother, to take steps to patch their code. Years after the discovery of Heartbleed, hundreds of thousands of devices and servers were still vulnerable.

```
buffer = OPENSSL_malloc(1 + 2 + payload + padding);
bp = buffer;
/* Enter response type, length and copy payload */
*bp++ = TLS1_HB_RESPONSE;
s2n(payload, bp);
memcpy(bp, pl, payload);
bp += payload;
/* Random padding */
RAND_pseudo_bytes(bp, padding);
```

The Heartbleed code and the logo developed for the vulnerability.

One of the primary lessons of Heartbleed was that open-source code like OpenSSL needed stronger institutional support for security, rather than just relying on volunteer efforts to find vulnerabilities. Years after its discovery, the Heartbleed vulnerability had been fixed on most devices and websites, but its legacy was an important one for thinking about our reliance on open-source software libraries and how we support the maintenance and security of these sorts of crucial, shared resources.

Part of what makes the Heartbleed vulnerability so striking—and part of the reason it led to so much soul-searching about how to do a better job with open-source code—is its simplicity. Much of it comes down to a straightforward idea about computer memory and a single line of code in the C programming language.

C is a very powerful language and also a terrifying (and often tedious) one to program in, partly because you can use it to directly manipulate a machine's memory—that is, tell a computer where to store certain bits in its memory or when to delete them and free up that memory for new bits of information. This can be a very useful and heady thing, but you also have to be constantly worrying about whether you've allocated and freed up memory correctly. That can be particularly hard to adjust to if you first learned to program with languages like Java or

Python that don't require thinking about computer memory. When you screw up memory management, one of two things can happen—either the program won't work and you'll go back to debugging to try to figure out where exactly you went wrong, or, even more frightening, it will work perfectly, and it will never even occur to you that you haven't handled the memory management properly.

The latter is what happened in the case of Heartbleed. When two computers are connected but not actively transferring data, one will occasionally send the other a "heartbeat" request to let the other know that they're still connected. When that happens, the computer that receives the request will respond, sending back the same piece of encrypted information it received, to confirm the connection. When that happens, the computer that sends the initial heartbeat request lets the receiver know exactly how long the piece of information they're sending is. But the OpenSSL software, crucially and terribly, forgot to verify that the length information for the heartbeat data was actually correct.

If that doesn't sound so horrifying to you, it's worth taking a look at the actual line of code behind Heartbleed that instructed the computer running it to copy the heartbeat request from a block of data of a particular size that is stored in memory to a new address:

```
memcpy(bp, pl, payload);
```

The `memcpy()` function performs the copying, the variable called "`bp`" designates where the information is going to be copied to, the "`pl`" variable designates the source where the information is currently, and the "`payload`" variable tells the program how much data should be copied from the source address or, in this case, how large the heartbeat is.

This works great so long as the computer sending the heartbeat request isn't lying about how much information it's sending (and requesting back). In fact, one of the reasons no one noticed Heartbleed for so long is that this code functions absolutely fine so long as no one involved has any malicious intentions. But if someone lies and claims to be sending a 60 KB heartbeat when they've actually sent a 10 KB one, then things deteriorate quickly. The receiver of that request would copy only 10 KB from the original heartbeat message but would use the variable of 60 KB that the sender provided them to determine how much information to send back to the sender. So they would send back the 10 KB message, but they would also send back another 50 KB from whatever was stored in its memory next to those 10 KB—which could be nothing interesting whatsoever... but could also be a confidential message or an important password or secret key.

The Heartbleed code is a chilling reminder of how hard it is to work in languages that afford programmers the power and responsibility of allocating and freeing up computer memory. The fix is relatively straightforward: check to make sure that the heartbeat message actually is as long as the sender says it is before responding. But remembering to think that way—to think like someone who would be deliberately trying to subvert the code, to think about the computer's memory and how it works, to do both of those things at the same time—is often the hard part.

The code that created Heartbleed and its consequences also underscores just how easy it can be to let small coding mistakes slip through the cracks when everything seems to be working fine, and just how massive the ramifications of those seemingly small mistakes can be. Finally, Heartbleed also reinforced for giant technology corporations just how dependent they are on open-source code and the crucial importance of finding ways to protect and secure code that is not the clear responsibility of any individual company or person. In the months that followed Heartbleed's discovery, 12 major technology firms, including Amazon, Cisco, Facebook, Google, IBM, and Microsoft, pledged to contribute more than $1 million per year to an endeavor dubbed the Core Infrastructure Initiative, intended to support open-source software like OpenSSL collectively.[3] That's not an enormous sum of money, especially given the size of the companies involved, but it's a great deal more than the roughly $2,000 in annual donations that OpenSSL received prior to Heartbleed to support its upkeep.[4]

The technical vulnerability behind Heartbleed was relatively straightforward. But the long-term solution for how we protect and maintain valuable open-source code like OpenSSL will be anything but simple. It will require the richest and most powerful companies in the tech sector to work together with their fiercest competitors, and to contribute time and money and expertise to going over lines of code that they did not write but must take responsibility for, nonetheless, or else no one will.

23
The Volkswagen Emissions Scandal
How Digital Systems Can Be Used to Cheat

Lee Vinsel

Word came down in September 2015: the prestigious German automaker Volkswagen had misled regulators and the public at large by using computer software to cheat air pollution tests. Researchers from West Virginia University found that one Volkswagen was spewing between 15 and 35 times the legal limit of nitrogen oxide, which can cause asthma, cancer, and heart attacks.[1]

VW fans took it as a slap in the face. The company had long touted its diesel vehicles as a green technology, meant in part to compete with Toyota's hybrid vehicles, most famously the Prius. In VW ads, sleek cars raced against backgrounds of glorious, mist-covered mountains, and mottos ("Like really clean diesel") hovered above the heads of happy, conscientious consumers.

A study out of MIT found that excess nitrogen oxide emissions from VW's cheater cars would lead to over 1,200 premature deaths in Europe alone—people gasping their last breaths from an asthma attack or grabbing their chests as their hearts explode from the fight to breathe.[2] That human toll is dramatic enough, especially if we rewrite the headline, "Corporation Toys with Computer Code and Murders over 1,000." But on another level, the Volkswagen emissions scandal shows us how software has become a weapon firms can use to circumvent regulations and the democratic values they embody. This opening for immorality—call it "innovation" in the service of crime—demands our attention.

In the United States, the regulatory context for Volkswagen's deceit was first set by the Clean Air Act Amendments of 1970. It was a law passed out of frustration and outright anger. Researchers in California had discovered that automobiles were responsible for Los Angeles's famous smog and other significant air pollution problems in the early 1950s. At that time, legislators, policymakers, and others saw air pollution as a local problem best handled by *local* decision-makers, not federal ones. And anyway, strong traditions of American liberalism made lawmakers extremely reticent to regulate industry.

Thus began a nearly 20-year period of legislative ineffectualness and industry resistance. Representatives at the local, state, and eventually federal levels passed toothless law after toothless law, and automakers swore to God on high that they were doing all they could to fix the problem. In 1969, the US federal government found that car companies were doing just the opposite: the Justice Department brought a lawsuit against the automakers under the Sherman Antitrust Act, arguing that the firms were using a research-sharing agreement perversely to guarantee that effective emissions controls never reached the market.[3]

The modern US environmental movement reached a crescendo in 1970. The first Earth Day was held. One of Ralph Nader's research groups published the book *Vanishing Air*, skewering automakers as purposeful polluters and clean air laws as feeble and hopeless. President Richard Nixon created the Environmental Protection Agency. And legislators, especially Senator Edmund Muskie, grew sick of industry inaction and, in disgust, passed the Clean Air Act Amendments of 1970, perhaps the strongest antipollution law created in

US history. (The law would receive further important amendments in 1990.)

The 1970 amendments were a clarion call for democratic accountability. They required automakers to reduce specific chemicals in emissions by 90 percent or face grave consequences, including large fines and even an order to halt production. Whether cars lived up to these requirements would be settled via tailpipe emissions tests (which many of us get done on an annual basis through state auto inspections).

The law obviously offered car companies strong incentives to get their acts together and control emissions. The catalytic converters that ride around on the bellies of our go-machines, tempting thieves with their precious metals, are the product of such efforts. But like all tests, for the unscrupulous, the law also created strong incentives to cheat. And cheat the companies did—indeed, almost immediately.

In 1972, staff members at the Environmental Protection Agency discovered that automakers were using mechanical devices to shut down emission controls under certain conditions.[4] Customers would buy these cars without ever knowing that their vehicle was usually out of compliance with pollution laws. To give one example of how this could work: federal rules mandated that tests be conducted in ambient air conditions between 68 and 86 degrees Fahrenheit. All automakers had to do is add devices that shut down emission controls when temperatures fell outside this range. EPA staff members dubbed such cheats "defeat devices," and in January 1973, William Ruckelshaus, the first EPA administrator, ordered automakers to remove all defeat devices from their vehicles within 90 days.

As the defeat devices demonstrate, automakers didn't require computers to commit crimes. But what many people do not realize is that many years before we began yelling at our vehicles, "Phone, CALL. MOM. No, no, no, I SAID . . . ," computers were successfully used to manage combustion. In fact, computers were first added to automobiles because of air pollution laws. In the late 1970s, automakers contracted with electronics firms, like Motorola and Intel, to install microprocessors in cars. Originally, the Big Three used computers only to control carburetion, and later fuel injection, to shape vehicle emissions. Then lawmakers—first in California in 1988 and then at the federal level in 1990—mandated that new cars include computerized onboard diagnostic tools, part of which became known as the "check engine light," perhaps the greatest icon ever of automotive computerization.

Computers opened up new opportunities for cheating pollution laws. If companies had been able to use mechanical devices to snooker tests in the early 1970s, computers gave them even finer control of how a vehicle would perform under a range of conditions. Never one to pass up on an attractive innovation, General Motors installed a computer chip on Cadillacs in the early 1990s that turned off emission controls whenever a car's air conditioning or heater was turned on.[5] The company knew full well that these systems were turned off during tests. In 1993, the EPA discovered that the Cadillacs produced three times the legal limit of carbon monoxide when the air was running. In the end, General Motors paid a $45 million fine for its deceit. But where GM innovated, others would follow.

Enter Volkswagen. As *New York Times* journalist Jack Ewing makes clear in his great book, *Faster, Higher, Farther: How One of the World's Largest Automakers Committed a Massive and Stunning Fraud*, in the 1990s and trailing into the first decade of the 2000s, Volkswagen faced enormous headwinds and even financial insolvency. Executives there embraced a cutthroat, "by all means necessary" approach to improving the company's position among competitors, including by placing their underlings under tremendous pressure.[6]

In 2007, a VW executive announced the ambitious plan to nearly double the number of vehicles it sold, from six million to ten. Increasing the number of cars sold in the United States was central to this vision, and in a bid to appeal to environmentalists, the firm's strategy came to focus on "clean diesel" cars.

There was a problem, though. Clean diesel cars weren't . . . clean. In 2007, a software engineer presented a PowerPoint at a meeting in VW's Research and Development building, showing how they could use computers to recognize when a car was being put through an emissions test. Emissions controls would function effectively under testing conditions, but not otherwise. When not being tested, so-called clean diesels could belch more nitrogen oxide than a new diesel tractor trailer.

As one academic article notes, today "a premium-class automobile can contain more than 70 control units and 100 million lines of code."[7] Engineers at Volkswagen changed code in its Engine Control Unit, which was manufactured by Bosch and was widely used in diesel passenger vehicles. The altered firmware used sensors to monitor when a vehicle was under test conditions and modified vehicle operations when it detected an emissions test. It was nearly a decade before these deceptive changes came to light.

The Volkswagen emissions scandal is a cautionary tale about how digital systems can be used to circumvent rules made under different technological realities. And of course, it applies to much more than just cars. When regulations depend on holding technological systems up to specific technical standards, firms can potentially use computers to spoof the test. Moreover, many companies, including the automakers, insist that the software in their products is covered by copyright and not open to public scrutiny. Where regulations require light, corporations demand opacity.

But the VW scandal raises questions that go far beyond corporate malfeasance. The communications scholar M. C. Forelle has found in her research that contemporary hot-rodders and tuners hack the computer systems in their cars to improve performance in ways that increase emissions beyond legal limits.[8] Then, when the owners need to take their car in for an annual inspection, they restore the software to factory settings—and the car passes the test perfectly.

Regulations are a key tool democracies use to require technologies to live up to widely shared values. In many areas, computerization forces us to ask, how can we ensure that the objects around us conform to the demands we place on them?

24

The Code That Brought a Language Online

Syeda Gulshan Ferdous Jana

August 17, 2005, was a day of terror in Bangladesh—and would also be pivotal toward bringing Bangla, the language of Bangladesh, online. It was a normal, busy office day in Dhaka, the capital, when one of the employees at my company received a call from his wife, saying that a small bomb had exploded in their neighborhood. Minutes later, another employee received a call about an explosion near his home, elsewhere in Dhaka. Chilling rumors started spreading about terror attacks by Islamist militants all over Bangladesh.

In fact, a small extremist group had exploded 500 bombs in nearly all of Bangladesh's 64 districts simultaneously. Word spread through mobile phones and one-to-one conversations, but it took several hours before TV news provided reliable updates and a full day before they became available in the online newspapers . People were frantic, searching for information to assess the risk for themselves, their families, and the country.

That was when we, the small team of the newly started social media company called "Somewhere in . . . ," saw the need for a citizen media: a real-time, dynamic online blog community. While my teammates discussed possible ways to meet that need, I realized there was a technical problem. At the time, there was no way for most people to write in Bangla online.

Bringing Bangla—the language of the liberal, Muslim-dominated country Bangladesh and the world's sixth largest language—online helped create a unique and vibrant blogosphere that allowed people to access information and, more importantly, express themselves. But it only happened because of some lines of code—or, more accurately, the people behind those lines of code. I'm proud to say that I am one of those people.

In December 2005, four months after the attacks, we launched the blog community platform somewhereinblog.net. It was the first of its kind in Bangladesh and the first to offer Bangla text entry, dynamic content, and user interaction. The platform and concept became highly influential, and both loved and hated. Our blog community offered a common front page displaying extracts of the latest posts from any of the bloggers, as well as an easy way to enter Bangla text, including phonetic and virtual Bangla keyboards.

This was revolutionary. Back in 2005, JavaScript had limited power, and it worked differently in all of the mainstream browsers like Internet Explorer, Firefox, and Opera. Bangla keyboard software for desktop use had become available via floppy discs a year earlier, but only experts could tweak blogging software like Blogspot and WordPress to accept and display this text.

Thanks to the Bangla phonetic keyboard developed in-house by Hasin Haider, our lead programmer, users could now type phonetic Bangla using any standard English keyboard and it would convert to Bangla on-screen. Our developers' main focus was to keep the code minimal, so that it would function properly across all browsers. This is how it worked: We captured keystrokes using the `keyDown` event, and then we calculated the equivalent phonetic representation of those keystrokes from an array. Finally, in the active text area, where the user was typing, we appended those Bangla characters at the position of the cursor, which was tricky at that time because every browser had to process the text differently. This was the most challenging—and important— part of the whole project.

We did not want to keep this powerful tool to ourselves. Our head developer released the code as open source, and we encouraged others to use it for free, to help cement Bangla's place online.

Our Bangla Blog Community concept became so popular that nearly 30 other similar platforms later sprang up. In total, I estimate that there were about 300,000 Bangla blogs at the peak of blogging in 2013, with about 50 percent on our platform. The addition of new platforms was something we supported and encouraged as part of our goal to enrich the Bangla blogosphere and help each blogger find a place to belong. Over the years, we saw many bloggers gain the courage to publish their own books after receiving encouragement from the community to develop their writing.

A typical Bangla blogger could best be characterized as 25 to 35 years of age, educated, working, and deeply engaged in supporting the country's development. Journalists, activists, students, politicians, teachers, writers, and others used the blog communities for research and opinion metering. Our bloggers wrote posts and took part in discussions in the comments about current events, social injustice, taboo-related social issues, women's empowerment, and corruption, but they also wrote beautiful poetry, colorful short stories, and insights into the lives of ordinary people. People did not blog for themselves, but for the whole blog community.

In collaboration and discussion with the bloggers, we created a set of rules and a "trust scale." Trust would be earned if writers obeyed the rules, and trust would be reduced for violations, with the trust score gradually rebuilding itself after it had been reduced. We had four categories on the trust scale. Blogs on the "green list" and "watch list" would appear on the blogger's personal page as well as the common front page, visible to all. For "watch list" blogs, moderators would review posts soon after posting for content control. We had a "red list" for blogs that had multiple or gross violation of the rules; these posts would not appear on the common front page until enough days had passed to restore the blog to a higher list status. Finally, we had a "blacklist" for permanent closure. For instance, one way to get on the blacklist was to post against the freedom and independence of the country—a criminal offence in Bangladesh. In this way we managed to establish blog etiquette and maintain freedom of expression with a minimum of moderator interference.

It was not easy to promote freedom of expression in Bangladesh. The nation had a very painful birth in 1971. The population was, and is, quite polarized in terms of political belonging and religious beliefs; many different groups claim to be the arbiters of truth.

We moderated posts based on the trust scale and also reviewed reports of abuse. Not much was removed, and often active and open participation in the debates by our team was sufficient to ensure that the conversation did not turn nasty. Still, our moderation was criticized many times. Even more often, we were criticized for *not* moderating. It is dangerous to moderate opinions, and we always strove to limit moderation to cases of hate speech and clear violations of the rules—rules that the blog community had been part of shaping. The problem was that neutrality is a vague term in a highly polarized nation.

The peak of Bangla blogging came in 2013, when—after years of demands from bloggers—a war-crime tribunal was launched, 42 years after the liberation of Bangladesh. Until then, the war criminals from the genocide carried out against the nation roamed freely through the country, some of them even serving as ministers in the Bangladesh Nationalist Party alliance government from 2001 to 2006. Upon hearing the news of a lenient verdict for one of the worst war criminals, Bangla bloggers called for a sit-in demonstration in Dhaka, demanding the highest punishment for the war criminals and a ban on Islamist politics and institutions. On the first night, 100 bloggers gathered at the busy Shahbag intersection in downtown Dhaka. The next night, perhaps 1,000 showed up. Within a few days, hundreds of thousands from all walks of life came to join the sit-in.[1] This was seen as a threat by the nationalistic parties and religious groups; many of those on trial were their leaders. Ten days into the uprising, an atheist blogger was killed,[2] and his "anti-Islam" writings were published by conservative media. In the years to come, bloggers were collectively tagged as atheists[3] and anti-Islam.[4] In response to a massive religious counter-demonstration against bloggers,[5] the progressive government jailed a handful of atheist bloggers, and it was reluctant to seek justice for the 25 or so bloggers and other prominent minority personalities who were killed.[6] At the same time, advertisers began to shy away from blogs, possibly under pressure from intelligence forces, restricting ad income for blogs and thereby making it hard to survive for most platforms. Somewhere In . . . is one of the few still operating, despite being blocked for a year by the authorities.[7]

For us, intimidation and threats came with the job. Our phones were tapped, our every move was shadowed, and we received numerous death threats through phone calls and Facebook. In the broad sense, these were confirmations and compliments indicating that the work we did had a real impact and influence on society. Freedom of expression is one of the most important pillars of a democracy and is an asset for any government to understand the sentiments of the people. Unfortunately, that opportunity was greatly missed in Bangladesh, which continues to lose ground on the press freedom index year by year. With the moral support of the Google Government Affairs and Public Policy team, in March 2013 we launched what is possibly the first Transparency Report of any Asian digital platform, half a year before Yahoo launched the first in the United States. Perhaps because of the transparency report, which would shed light on government interference, Somewhere In . . . has experienced significantly less intimidation since 2013.[8]

In retrospect, these lines of code affected the lives of millions, mostly in a positive way, but it put others in danger for their activism or just for their beliefs. Still, the feeling of empowering people is the highest reward, which we have reaped for all of these years.

25

Telegram

The Platform That Became "the Internet" in Iran

Mahsa Alimardani and Afsaneh Rigot

Since the introduction of the Internet to Iran in the late 1990s, Iranian authorities have seen it as a threat to national security and have been preoccupied with finding innovative ways to control it.[1] Iran's head of state, the Supreme Leader Ayatollah Khamenei, regularly says in his sermons that the Internet is a space that Iran's enemies try to infiltrate with "psychological warfare."[2] This is the pretext for all the censorship and information controls the Islamic Republic tries to apply online—that they are trying to protect the country from "Western" attacks that turn Iranians against the Islamic Republic.

Over the past three decades, there have been countless examples of digital censorship and surveillance in Iran. But none better demonstrates the peculiarities of Iran's Internet conditions than the messaging and social media application Telegram and how it has affected Iranian society and politics. Few technologies have generated as much concern among Iran's leaders as Telegram, which has functioned as both a friend and threat to the regime—and to the Iranian Internet users trying to evade the government's censorship, targeting, and monitoring efforts. At the heart of Telegram in Iran is a story of Iranian ingenuity, as users found ways to resist the confines of authoritarian controls.

Telegram was launched in 2013 by Pavel Durov and his brother Nikolai. When they launched the platform, they promised it would be both "secure and fun."[3] To fulfill the "fun" part, they aimed to create a community around the technology. As part of that, they left their API open, so other developers could make their own third-party versions or add additional features on the application.[4] (An API is an Application Programming Interface, and in keeping it open, Pavel and Nikolai allowed outside software, operating systems, and microservices to interact with the app itself.) This allowed developers to create a whole host of new features that made the platform extremely popular among Iranians, like custom Persian stickers or bots that could generate searches and queries for you in an instant. The latter was of particular use to Iranians: they could use Telegram bots to quickly and easily access information, without having to turn to browsers or search queries that could be extremely cumbersome on Iran's famously slow and brittle Internet. The bots were all the more useful when people wanted to engage in anything that would be taboo on the Iranian Internet—like finding pornography or researching sensitive topics like religion. However, the same open API would later allow developers affiliated with Iranian authorities to create their own "forks," or third-party versions of the application that would eventually compromise the data and privacy of users in Iran.[5]

In September 2015, Telegram launched a new feature that changed Iran's communication and social media landscape profoundly: channels. Telegram channels are like forums, except that administrators are able to post messages, but other users are not. Channels allow administrators to broadcast messages to an unlimited number of subscribers.[6] Soon it became impossible to have any sort of media presence in Iran without hosting a Telegram channel. Media organizations whose websites were censored in Iran, such as BBC Persian, could upload entire articles or broadcasts to their Telegram channels, where they were followed and viewed by millions of users inside Iran. Telegram allowed for uncensored transfer of large files, media, and communications to millions of Iranian users with unprecedented speed and ease. (Competitors like WhatsApp and Viber didn't accommodate as many types of files, for instance.) By 2017, Iran's Ministry of Information Communications and Technology announced that Telegram use inside of Iran made up 15 percent of all of Iran's international Internet traffic.[7]

As early as 2015, Iran's politicians, officials, and state media organs started to express concern over the free flow of unfiltered information Telegram provided to users. Iran's filtering committee—a multiagency body that recommends what should be censored—repeatedly deliberated blocking Telegram.[8] At times, Telegram even cooperated with the government guidelines, such as when it censored porn bots only in Iran;[9] removed a channel heavily involved in protest mobilization during Iran's 2017–18 nationwide protests;[10] and, finally and most critically, began hosting their content delivery network servers in Iran in 2017.[11] Telegram was officially banned in April 2018 on national security grounds,[12] though the real reason was presumably the role it played during the 2017–18 antigovernment protests.[13] At that point it had effectively become "the Internet in Iran."[14] Even though the government has blocked it, as of late 2021 Telegram remains one of the most used applications in Iran, thanks to circumvention tools like virtual private networks.[15]

Telegram came to prominence under the guise (or branding) of a "secure messenger" that used end-to-end encryption. But from the start, there was confusion as to how secure Telegram actually was. By default, the messenger relied on plain text messaging that was not encrypted—users had to turn it on in private chats. Some of Telegram's most popular features, such as channels and bots, didn't use any security protocols.[16] There were also questions from leading security researchers about the effectiveness of Telegram's end-to-end encryption protocol.[17]

But Western security experts' discussions about these flaws were frequently misleading. In the West, conversations about the security and safety of messaging platforms revolve around the strength of the tool's encryption infrastructure and its han-

A map of the National Information Network, effectively an Iranian Internet that is hosted domestically, with restricted access to outside content. The Iranian government is able to impose shutdowns, in which key services—banking, e-commerce, messaging and streaming apps—remain intact, but access to international media is denied.

dling of user data. But in Iran, law enforcement doesn't bother cracking encryption protocols to identify and arrest people—it's more straightforward than that. Instead, they weaponize the app itself by infiltrating groups and channels (either through fake accounts, social engineering, or similar methods) and then identifying admins and users.

These approaches are particularly dangerous for people who are already at risk, such as members of queer communities. LGBTQ+ Iranians are marginalized and face legal threats for living their lives, while having no access to queer platforms like Grindr or Manjam. As the center of online communication in Iran, Telegram quickly found use as a queer dating app, information hub, and community connector. This in part was due to its perceived security.[18] But that security was largely an illusion. In 2016, the website for the Revolutionary Guards' cyber unit warned that "sodomites are parading on Telegram in the shadow of officials' negligence."[19] They published screenshots of LGBTQ Telegram groups allegedly used for dating, including profiles of people from these channels. The article also warned that some of these groups were trying to "promote" homosexuality while working to secure "recognition and rights for homosexuals." In another report in 2017, a prosecutor from the city of Ardebil told the Tasnim News Agency that "six administrators of Telegram channels have been arrested in Ardebil" over promoting sodomy and immorality.[20] Many queer-focused groups were infiltrated by law enforcement personnel, who gained access to the groups' content. Group and channel admins were identified, and many reported getting warnings and subpoenas sent to them after they were identified. Others were arrested.[21]

Since the 2018 censorship of Telegram and queer-focused dating apps, the community has had to revert to using even riskier online tools, such as forks like Telegram Talaei and Hotgram. In order to access blocked dating applications, messaging, or social platforms like Telegram, queer people have had to rely on virtual private network and proxy services. But reports suggest that some popular VPN and proxy services are tied to Iran's authorities and have been revealed to have recorded user data—major security risks to groups already disproportionately in danger.[22]

Today, Telegram's role in Iran has faded somewhat from its peak. According to the most recent Internet statistics, WhatsApp[23] is now the most used social application in Iran.

During its heyday, Telegram became an all-encompassing hub of commercial, social, and political life, facilitating some of the most fragile and intimate connections for people in Iran, including for some of the most marginalized. The impact of Telegram's technology wasn't necessarily due to its "code" or infrastructure. It came from the way that Telegram filled in the gaps of Iran's heavily censored Internet—but those gaps could also be traps. Telegram could be both a lifeline, when used ingeniously by the people of Iran—or a prison sentence, when weaponized by law enforcement.

The authors would like to thank Sayeh Isfahani for the research contributions and translations as part of the bigger research project Afsaneh Rigot is conducting on Queer communities and the impact of tech in MENA.

26

Encoding Gender

Meredith Broussard

Jonathan Ferguson, a 40-year-old technical writer at the Ministry of Supply in London, made UK headlines in 1958 when he formally announced his gender transition. "His birth registration has been amended from 'female' to 'male' and his new Christian name inserted into the register," reported the UK's *Daily Telegraph and Morning Post*. This quote, which I read in a paper by scholar Mar Hicks called "Hacking the Cis-team: Transgender Citizens and the Early Digital State,"[1] has stuck with me because it suggests ease.

In 1958, Ferguson's interaction with the state around his transition involved changing a form and issuing a card. It's not dissimilar to the way that a name or gender change is handled today—though the paperwork and expense now are far more substantial. From a sociotechnical perspective, Ferguson's card resonates because it reminds me that 1950s culture still governs our lives today through the design of large bureaucratic computer systems. This is particularly true when it comes to interacting with the government. Despite advances in concepts of gender, and advances in LGBTQIA+ rights, most computer systems still encode gender as a binary value that can't be changed. The next frontier in gender rights is inside databases.

Computing as we know it today started in 1951, when the Census Bureau started running the first commercially produced digital computer, UNIVAC. Back then, gender was generally considered fixed. If you filled out a paper form, it asked for your name and offered you two choices for gender: male or female. You could pick one. Computer programmers used those same paper forms to design computer databases, and when you designed a record for a database, it looked something like this:

```
Firstname
Lastname
Gender (M/F)
Address 1
Address 2
Zip
```

This was how computer database design was taught through the 1990s, when I learned programming. Now, we have a more comprehensive understanding of gender, and an increasing number of companies are embracing inclusive design principles that allow users to self-identify in databases as nonbinary, transgender, genderqueer, and other terms that encompass a range of LGBTQIA+ identities. However, there are artifacts and idiosyncrasies inside computational systems that serve as barriers to implementing truly inclusive design. Most of these problems come from the way that 1950s US and UK social perspectives informed how computer schemas were created.

Most of the intellectual history and the dominant social attitudes in the field of computer science can be found in a single, sprawling database published by the Association for Computing Machinery (ACM). The earliest mention of gender in the ACM Digital Library comes in 1958, in "The Role of the Digital Computer in Mechanical Translation of Languages."[2] It had to do with translation and pronoun matching in translation—and for the next 20 years, all mentions of gender have to do with translation. In other words, even though sweeping social change happened in the 1960s and 1970s, academic computer science (and for the most part, the computer industry) pointedly ignored the topic of gender except to think about how a computer might accurately translate gendered pronouns from one language to another.

When same-sex marriage was legalized in the United States, it required changing thousands of database schemas. The database redesign process was informally called Y2gay. Most databases were set up to only allow marriages between men and women; changing the law required changing those databases to comply. The name Y2gay is a reference to the Y2K problem, which occurred in the 1990s when people realized that most databases and code stored the date as two digits, using an implied prefix of 19. Changing over to the year 2000 was going to screw up an awful lot of code.

Facebook, which began as a kind of "Hot or Not?" for male undergraduates to rate women, was among the first social media companies to allow users to change their names and gender identity, both of which are required at signup. Although its software allows users to self-identify on their profiles as one of more than 50 different genders, the way the system actually stores the data is that each user is recorded (and sold to advertisers) as male, female, or null.

The reason for this has to do with both hegemonic heteronormativity and math. Everything you do on a computer is secretly math. Many computer programs exist to slot people into neat categories in order to do data analysis, creating a tension between the messiness of the "real" world and people's shifting identities. This is most obvious when it comes to the gender binary and binary representation in computer systems.

You know the gender binary: the idea that there are two genders, male or female. Binary code is also the system that powers computers. In a binary numeral system, there are only two numbers: 0 and 1. The numbers 0 through 4 look like this in binary:

```
0: 0
1: 1
2: 10
3: 11
4: 100
```

Computers are powered by electricity, and the way they work is that there is a transistor, a kind of gate, through which electricity flows. If the gate is closed, electricity flows through, and that is represented by a 1. If the gate is open, there is no electricity, and that is represented by a 0. I'm simplifying it dramatically, which will enrage a certain kind of nerd, but here's the gist: this unit of information, a 1 or 0, is called a bit. There are 8 bits in a byte (except when there aren't), and a million bytes in a megabyte. This is how we talk about memory space in computing. Programmers are always thinking about how much memory space a program takes up or how much data can be stored because space on a computer is finite.

Different arrangements of bits can be mapped to letters or numbers. In the United States, the most common mapping is called ASCII. In ASCII, the letter A is represented as 01000001. My first name, Meredith, looks like this in the ASCII version of binary representation:

```
01001101 01100101 01110010 01100101
01100100 01101001 01110100 01101000
```

When a computer stores information about the world, we call that information data. Data is stored inside a database. In a database, every piece of data has a type, and usually the rules for that type are very strict. In the very simplest form, we can think of data as being of three types: letters, numbers, or binary (0 or 1) values. A binary value is often referred to as a Boolean, named after a nineteenth-century guy named Charles Boole who invented a system of logic that only uses 1s and 0s. If you want to use data in a computer program, you feed that data to a thing in the program called a variable. Variables also have types, and those types are strictly governed by the rules and vocabulary of a specific programming language. While their rules may be different, all programming languages have the same essential forms, meaning that they all on some level translate keyboard strokes, mouse movements, variables, data, and so on into binary. This is why a computer can't work without power.

So: in order to store data, we have to declare variables of a certain type inside a database. Speaking loosely, the types are string (meaning text, as in a string of letters), or number, or binary (aka Boolean). Boolean variables are used when a value is true or false, and are represented as `1` (true) or `0` (false). That looks something like this:

```
Firstname  [string]
Lastname   [string]
Gender     [Boolean]
Address 1  [string]
Address 2  [string]
Zip        [number]
```

In our sample database record above, we'd have to make certain decisions about each field. What type of data goes into each field? How large does each field need to be to hold the intended data? Who can enter the data? Who can change the data? Under what circumstances can the data be changed? Which fields can be edited, and which are fixed? Someone decides these factors every single time a database is created. But though they may seem mundane, these decisions can be consequential. This is not about math, but about implementing human social values inside a mathematical system. The question becomes: Whose values are encoded in the system?

In a paper called "The Misgendering Machines: Trans/HCI Implications of Automatic Gender Recognition,"[3] scholar Os Keyes read and analyzed all of the top academic work on automated gender recognition from 1995 to 2017 and found that the overall assumption is that gender is binary, immutable, and/or physiological.

Rigid assumptions about cisgender heteronormativity are amplified inside databases because even something as seemingly small as choosing free text entry versus a dropdown has implications. A letter occupies more bits and thus takes up more memory space. Today, it's easy to ignore memory concerns, but until the late 1990s computer memory was expensive. Using a Boolean variable is extremely efficient. A `0` or `1` takes up less space than `01001101` (M) or `01000110` (F).

If you are designing code for maximum speed and efficiency using a minimum of memory space, you try to give users as few opportunities as possible to screw up the program with bad data entry. A Boolean for gender, rather than a free text entry field, gives you an incremental gain in efficiency. It also conforms to a certain normative aesthetic known as "elegant code."

That aesthetic of "elegant code" is specifically exclusionary to someone like Zemí Yukiyú Atabey, who identifies as genderqueer and nonbinary. Atabey's pronouns are ze (where is ze?) zeí (zeí isn't coming today, sorry) and zem (I don't have the tickets, I gave them to zem). "As a nonbinary person, there is no option most of the time," ze says of entering personal information in databases. Microsoft Word, the program I used to compose this essay, marked all of Atabey's pronouns with the red squiggly underline, meaning that the people at Microsoft who wrote Word do not recognize Atabey's pronouns as acceptable English words, even though the genderqueer community has been suggesting ze and hir as pronouns for at least 20 years.

I met Atabey when ze was a graduate student. New York University, Atabey's school (and my employer), is among the most progressive universities when it comes to gender identity. Students can change their gender identity in Albert, the student information system.

In the Albert documentation, a distinction is made between legal sex ("a person's sex as currently indicated on a birth certificate, passport, or other official document") and gender identity ("the gender with which a person identifies"). It notes, "One's gender identity may or may not differ from one's legal sex which is assigned at birth based on biological characteristics."

Making this change was a complex matter. Most university student information systems—the core that everything feeds off of—were set up in the 1960s and haven't been overhauled. Changing them is complicated and expensive. It is absolutely possible to update university systems to encompass gender identity, just as it is possible to update financial systems, insurance systems, health systems, government systems, and every other system that relies on legacy design. It's a matter of will and funding.

The will to change is being achieved at long last through lawsuits and legislation. As of this writing, a third gender option, X, on official state documents has been mandated in 19 US states and in Washington, DC. Unfortunately, change is slow, and adoption is spotty. Even after the legislation is passed, more lawsuits have followed to force compliance. In March 2021, a group of New Yorkers sued because the databases that control access to Medicaid, food stamps, and other public assistance don't include the X option. "Any time I need something as simple as food or to make a doctor's appointment, I basically am forced to misgender myself, to be misgendered. And this takes a toll," coplaintiff Jaime Mitchell said.[4] Mitchell, who is nonbinary, was able to get their birth certificate with an X but was forced to declare themself male or female in order to get public benefits.

Lines of code can change the world, absolutely. In celebrating that fact, we need to also look at the way lines of code make culture incarnate and make social change much harder. Computer systems are not just mathematical, they are sociotechnical, and they need to be extensively updated on a regular basis. Just like humans.

Acknowledgments

This book would not have been possible without Slate's Lowen Liu. Lowen came up with the idea for the 2019 *Slate* article that inspired this book, "The Lines of Code That Changed Everything," which was made up of 33 blurbs about particularly significant lines of code. But the article also could not have happened without other *Slate* folks, some of whom are now at other publications: Jonathan Fischer, April Glaser, Jared Hohlt, Jane C. Hu, Lisa Larson-Walker, Aaron Mak, Susan Matthews, Molly Olmstead, Faith Smith, and Megan Wiegand all wrote blurbs, helped wrangle and edit, directed the art, and/or kept me from losing my mind. Elena Botella, who had only just started at *Slate*, not only wrote several entries—one of which she expanded on for this book—but fact-checked the code and kept the project in motion. Thank you, all!

The original *Slate* article also contained blurbs by an absolutely brilliant group of contributors: Arthur Daemmrich, Charles Duan, James Grimmelmann, Thomas Haigh, Amanda Levendowski, David S. Levine, Charlton McIlwain, Chris Noessel, Rusty D. Pickens, P. W. Singer, Nikki Stevens, Ellen Stofan, Clive Thompson, Lav Varshney, Sara Wachter-Boettcher, Ari Ezra Waldman, Josephine Wolff, Jamie Zawinski, and Ethan Zuckerman. Your work and your passion for it inspires me. Some of those write-ups have been expanded into essays here, but I wish we could have included all of them.

PUP's Hallie Stebbins saw the original *Slate* article and asked me, "Have you thought about making it into a book?" I hadn't thought about it, and it seemed an insurmountable challenge—but here we are. I definitely couldn't have done it with any other editor. Thank you also to Kelly Chudler for the absolutely inspiring illustrations, Michele Rosen for the fantastic copy editing, and Natalie Baan for production—it was a joy to work with all of you.

I am grateful to all of the folks at (or formerly of) New America and Arizona State University who have supported Future Tense for more than a decade, including but very much not limited to Michael Crow, Jim O'Brien, Ed Finn, Joey Eschrich, Anne-Marie Slaughter, Steve Coll, Fuzz Hogan, Alison Yost, and many, many more. In particular, Andrés Martinez, my longtime boss and friend, has offered me tremendous support in this project and so many others—thank you for not only encouraging me to try new things, but always helping me figure out how to make it work, even if it requires more phone calls than I would prefer.

And thank you above all to Nick, Brian, Kirstin, Fiona, Neve, my late parents (my mother would have been thrilled to see it officially noted in a book that she spotted the Y2K problem long before 1999), Bob, Diane, and, of course, Chris. Chris, I am the luckiest.

Notes

Preface

1 For more details about RSA, see James Grimmelmann's essay, "Wear This Code, Go to Jail," in this book (chapter 16).

Introduction

1 Steve Lohr, "John W. Backus, 82, Fortran Developer, Dies," *New York Times*, March 3, 2007, last accessed January 20, 2022, https://www.nytimes.com/2007/03/20/business/20backus.html.
2 Lohr, "John Backus."
3 "John Backus Quotes," quotefancy, last accessed January 20, 2022, https://quotefancy.com/quote/1745630/-John-Backus-I-myself-have-had-many-failures-and-I-ve-learned-that-if-you-are-not-failing.
4 "The automatic operations of System 1 generate surprisingly complex patterns of ideas, but only the slower System 2 can construct thoughts in an orderly series of steps." Daniel Kahneman, *Thinking Fast and Slow* (New York: Farrar, Straus & Giroux, 2013), 23.
5 "Unix," Wikipedia Foundation, last edited November 21, 2021, 05:59 (UTC), https://en.wikipedia.org/wiki/Unix.
6 "Bell Laboratories," Britannica, last accessed January 20, 2021, https://www.britannica.com/topic/Bell-Laboratories.
7 "The C Programming Language," Wikipedia Foundation, last edited January 8, 2022, 19:13 (UTC), https://en.wikipedia.org/wiki/The_C_Programming_Language.
8 Shishir Prasad, "Brian Kernighan: No One Thought C Would Become So Big," *Forbes India*, updated February 27, 2014, https://www.forbesindia.com/interview/special/brian-kernighan-no-one-thought-c-would-become-so-big/29982/1.
9 "History of the Internet," Wikipedia Foundation, last edited January 15, 2022, 17:58 (UTC), https://en.wikipedia.org/wiki/History_of_the_Internet.
10 "The Risks Digest," Vol. 32, Association for Computing Machinery, Committee on Computers and Public Policy, Peter G. Neumann, moderator, January 16, 2020, https://catless.ncl.ac.uk/Risks/32/index.
11 Bruce Schneier, "Schneier on Security," *Cryptogram Newsletter*, last accessed January 20, 2022, https://www.schneier.com/crypto-gram/.
12 Nicole Perlroth, *They Tell Me This Is the Way the World Ends: The Cyberweapon Arms Race* (New York: Bloomsbury, 2021), quoted in Jill Lepore, "The Next Cyberattack Is Already Under Way," *New Yorker*, February 8, 2021, https://www.newyorker.com/magazine/2021/02/08/the-next-cyberattack-is-already-under-way.
13 Raphael Satter, "US investigators probing breach at code testing company Codecov," *Reuters*, April 16, 2021, https://www.reuters.com/technology/us-investigators-probing-breach-san-francisco-code-testing-company-firm-2021-04-16/.

14 Andy Greenberg, "The Full Story of the Stunning RSA Hack Can Finally Be Told," *Wired*, May 20, 2021, https://www.wired.com/story/the-full-story-of-the-stunning-rsa-hack-can-finally-be-told/,

15 Greenberg, "Full Story."

16 Perlroth, *They Tell Me*.

17 "FORTRAN: The Pioneering Programming Language," IBM 100, last accessed January 20, 2022, https://www.ibm.com/ibm/history/ibm100/us/en/icons/fortran/.

18 Sarah Pruitt, "What Went Wrong on Apollo 13?," History.com, updated April 13, 2020, https://www.history.com/news/apollo-13-what-went-wrong.

19 "Apollo 13," Wikipedia Foundation, edited January 20, 2022, 19:40 (UTC), https://en.wikipedia.org/wiki/Apollo_13.

20 Hermann Dür and Eric Jones, "Building an Apollo 13 LiOH Canister Adapter," *Apollo 13 Lunar Surface Journal*, last revised 30 September 2012, https://www.hq.nasa.gov/alsj/a13/a13_LIOH_Adapter.html.

21 Pruitt, "What Went Wrong."

22 Nicole Perlroth, "Daniel Kaminsky, Internet Security Savior, Dies at 42," *New York Times*, last updated April 28, 2021, https://www.nytimes.com/2021/04/27/technology/daniel-kaminsky-dead.html.

23 Perlroth, "Daniel Kaminsky."

24 Robert McMillan, "Kaminsky: Many Ways to Attack with DNS," *Infoworld*, August 6, 2008, https://www.infoworld.com/article/2652414/kaminsky--many-ways-to-attack-with-dns.html.

25 Daniel Kaminsky, "The Hidden Architecture of Our Time: Why This Internet Worked, How We Could Lose It" (keynote address), 2016 Black Hat Hacker Conference, August 16, 2016, YouTube video, last accessed January 21, 2022, https://www.youtube.com/watch?v=RAGwol-keXM. His statement about the "DNS mess" occurs at approximately 24:35 of the video.

26 Perlroth, "Daniel Kaminsky."

27 Apollo launched in April 1970. "Apollo 13," Wikipedia. The Mars Orbiter launched on December 11, 1998. "Mars Climate Orbiter," Wikipedia Foundation, last edited on 8 January 2022, 12:19 (UTC), https://en.wikipedia.org/wiki/Mars_Climate_Orbiter.

28 *Launch Mishaps—Early Rocket Failures at Cape Canaveral Air Force Station*, produced by Murphy Wardman, narrated by Ken Flemming (US Space Walk of Fame Foundation DBA The American Space Museum), 2020, YouTube video accessed June 14, 2020, https://www.youtube.com/watch?v=ytaEtl2Nka4.

29 "Mars 2020 Mission, Perseverance Rover Launch," NASA Science, Mars 2020 Perseverance Rover, last accessed January 20, 2022, https://mars.nasa.gov/mars2020/timeline/launch/.

30 "Mars Timeline Landing Summary," NASA Science, Mars 2020 Perseverance Rover, archived February 18, 2021, https://mars.nasa.gov/mars2020/timeline/landing/.

31 Kenneth Chang, "NASA's Mars Helicopter Flies Again and Gets a New Mission," *New York Times*, April 30, 2021, updated May 10, 2021, https://www.nytimes.com/2021/04/30/science/mars-helicopter-nasa.html.

32 "NASA's Perseverance Rover Is Midway to Mars," NASA, Perseverance Mars Rover, October 27, 2020, https://www.nasa.gov/feature/jpl/nasas-perseverance-rover-is-midway-to-mars
33 Email from David C. Agle, Perseverance/Mars Helicopter/Juno Lead, NASA, received October 7, 2021.
34 "Ingenuity (helicopter)," Wikipedia Foundation, last edited on 21 January 2022, 01:45 (UTC), https://en.wikipedia.org/wiki/Ingenuity_(helicopter).

Chapter 1 / The First Line of Code

1 "History of Power: The Evolution of the Electric Generation Industry," *Power*, last updated December 22, 2020, https://www.powermag.com/history-of-power-the-evolution-of-the-electric-generation-industry/; Abbot Payson Usher, *A History of Mechanical Inventions*, Revised Edition. (1954; New York: Dover Publications, 1988), 289–90.
2 Melinda Watt, "Textile Production in Europe: Silk, 1600–1800," *Heilbrunn Timeline of Art History*. The Met, October 2003, https://www.metmuseum.org/toah/hd/txt_s/hd_txt_s.htm/; Usher, *History of Mechanical Inventions*, 289.
3 Usher, *History of Mechanical Inventions*, 289.
4 Usher, *History of Mechanical Inventions*, 292–93.
5 Franklin Allen, *The Silk Industry of The World at the Opening of the Twentieth Century* (New York: Silk Association of America, 1904), 20, https://hdl.handle.net/2027/umn.31951000946672j; Edgar C. Smith, "The Centenary of Napoleon," *Nature* (May 5, 1921): 20, https://www.nature.com/articles/107302a0.pdf/.
6 Alfred Barlow, *The History and Principles of Weaving by Hand and by Power*, reprinted with considerable additions from "Engineering," with a chapter on lace-making machinery reprinted from "Journal of the Society of Arts" (London: Sampson Low, Marston, Searle & Rivington, 1878), 141.
7 Benjamin Wooley, *The Bride of Science: Romance, Reason and Byron's Daughter* (London: Macmillan, 1999). 257, 268.
8 Stan Augarten, *Bit by Bit: An Illustrated History of Computers* (New York: Ticknor and Fields, 1984), chap. 2.8, http://ds-wordpress.haverford.edu/bitbybit/bit-by-bit-contents/chapter-two/the-analytical-engines-machinery/; *Encyclopedia Britannica Online,* "Analytical Engine," accessed April 18, 2021, https://www.britannica.com/technology/Analytical-Engine.
9 Alan Turing, "Computing Machinery and Intelligence," *Mind*, no. 59 (1950): 433–60.
10 "Herman Hollerith," Columbia University Computing History, Columbia University, last modified January 21, 2021, http://www.columbia.edu/cu/computinghistory/hollerith.html/; "Herman Hollerith," United States Census Bureau, last modified February 3, 2021. https://www.census.gov/history/www/census_then_now/notable_alumni/herman_hollerith.html/
11 "The Punched Card Tabulator," Icons of Progress, IBM, accessed March 19, 2021. https://www.ibm.com/ibm/history/ibm100/us/en/icons/tabulator/

12 Teun Kotsier, "On the Prehistory of Programmable Machines: Musical Automata, Looms, Calculators," *Mechanism and Machine Theory* 36, no. 5 (May 2001): 590; David Suisman, "Sound, Knowledge, and the 'Immanence of Human Failure:' Rethinking Musical Mechanization through the Phonograph, the Player-Piano, and the Piano," *Social Text* 28, no. 1 (Spring 2010): 19, https://doi.org/10.1215/01642472-2009-058; Thomas W. Patteson, "Player Piano," *Oxford Handbooks Online*, https://doi.org/10.1093/oxfordhb/9780199935321.013.16.

Chapter 2 / Monte Carlo Algorithms

1 Electronic Numerical Integrator and Computer
2 Iosif Shklovsky, *Five Billion Vodka Bottles to the Moon*, trans. Harold Zirin & Mary Fleming Zirin (New York: W. W. Norton, 1991).
3 Roger Eckhardt, "Stan Ulam, John von Neumann, and the Monte Carlo Method," *Los Alamos Science* Special Issue (1987).
4 Marshall Rosenbluth, "Genesis of the Monte Carlo Algorithm for Statistical Mechanics," AIP Conference Proceedings 690, no. 22 (2003).
5 Mathematical Analyzer Numerical Integrator and Automatic Computer Model
6 Using the Lennard-Jones potential.

Chapter 3 / Jean Sammet and the Code That Runs the World

1 With gratitude to her friend, Kathy Kleiman, for this turn of phrase.
2 Thomas J. Bergin, "Jean Sammet: Programming Language Contributor and Historian, and ACM President," *IEEE Annals of the History of Computing* 31, no. 1 (January-March 2009): 76–85.
3 "Ten Minutes with Jean Sammet '48," Mount Holyoke College Alumnae Association, last modified November 5, 2013, https://alumnae.mtholyoke.edu/blog/sammet/.
4 Thomas J. Bergin, "Jean Sammet."
5 "Ten Minutes With Jean Sammet '48."
6 Steve Lohr, *Go To: The Story of the Math Majors, Bridge Players, Engineers, Chess Wizards, Maverick Scientists, and Iconoclasts—the Programmers Who Created the Software Revolution* (New York: Basic Books, 2008), 47.
7 John Backus, "Programing in America in the 1950s: Some Personal Impressions," in *A History of Computing in the 20th Century*, eds. N. Metropolis, J. Howlett, and Gian-Carlo Rota (New York: Academic Press, 1980), 127.
8 Betty Holberton, "COBOL Session: Transcript of Discussant's Remarks," in *History of Programming Languages*, ed. Richard L. Wexelblat (New York: Academic Press, 1981), 262.
9 Dr. Tim Bergin, interview with author, March 26, 2021.
10 The acronym appears to have formed spontaneously. As IBM's R. W. Bemer—an advisor to the CODASYL committee—wrote in 1971, "we can't find a single indi-

vidual who admits to coining the acronym." In R. W. Bemer, "A View of the History of COBOL," *Honeywell Computer Journal* 5, no. 3 (1971): 132.

11 Edsger W. Dijkstra, *Selected Writings on Computing: A Personal Perspective* (New York: Springer-Verlag, 1982), 130.

12 Jeffrey R. Yost, "Where Dinosaurs Roam and Programmers Play: Reflections on Infrastructure, Maintenance, and Inequality," *Interfaces: Essays and Reviews on Computing and Culture*, May 2020, https://cse.umn.edu/cbi/interfaces.

13 Jean E. Sammet, "The Use of English as a Programming Language." *Communications of the ACM* 9, no. 3 (March 1966), 228–30.

Chapter 4 / Spacewar

1 Spacewar! has appeared with or without an exclamation mark and in all caps or in title case throughout its history. The original PDP-1 did not include an exclamation mark in its character set, and DEC marketing materials likewise did not include the mark. This essay follows the convention adopted by the Special Interest Group on Computers, Information, and Society (SIGCIS) of the Society for the History of Technology to not include the mark.

2 "PDP-1 Handbook," Digital Equipment Corporation, 1963, https://www.computerhistory.org/pdp-1/_media/pdf/DEC.pdp_1.1963.102636240.pdf

3 Oral history interview with Stephen Russell, January 8, 2017, Video Game Pioneers Oral History Collection, Smithsonian National Museum of American History Archives Center, 18, https://sova.si.edu/record/NMAH.AC.1498

4 Oral history interview with Daniel Edwards, November 30, 2018, Video Game Pioneers Oral History Collection, Smithsonian National Museum of American History Archives Center, 6, https://sova.si.edu/record/NMAH.AC.1498

5 J. Martin Graetz, "The Origin of Spacewar," *Creative Computing* 7, no. 8 (1981): 56–67.

6 Oral history interview with Stephen Piner, November 29, 2018, Video Game Pioneers Oral History Collection, Smithsonian National Museum of American History Archives Center, https://sova.si.edu/record/NMAH.AC.1498

7 J. Martin Graetz, "The Origin of Spacewar," *Creative Computing* 7, no. 8 (1981): 56–67.

8 The technical details are spelled out in: Norbert Landsteiner, "Inside Spacewar! Intermission: Digging up the Minskytron Hyperspace," June 5, 2014, https://www.masswerk.at/spacewar/inside/insidespacewar-minskytron-hyperspace.html

9 "PDP-1 Computer and Spacewar," Digital Equipment Corporation, 1963, https://www.masswerk.at/spacewar/pdp-1-computer-and-spacewar.html

10 Bill Pitts, "Letter to Stanford Magazine," *Stanford Magazine* (April 30, 2012), https://stanfordmag.org/contents/bill-pitts-68.

11 Mark Wolf, "Introduction," in *Before the Crash: Early Video Game History*, ed. Mark Wolf (Detroit: Wayne State University Press, 2012), 1–8.

12 For a broader perspective on video game longevity, see Raiford Guins, *Game After: A Cultural Study of Video Game Afterlife* (Cambridge, MA: MIT Press, 2014).

13 Norbert Landsteiner, "Inside Spacewar!: A Software Archeological Approach to the First Video Game," mass.werk (June 2014–March 2016), https://www.masswerk.at/spacewar/inside/

14 Robert Garner, "Restoring and Demonstrating 1960s Vintage Computers at the Computer History Museum," in *Making IT Work*, ed. Martin Campbell-Kelly (London: British Computer Society, 2017): 54–67, https://www.computerconservationsociety.org/miw/Proc%20MIW%202017.pdf.

15 Oral history interview with Stephen Russell, January 8, 2017, Video Game Pioneers Oral History Collection, Smithsonian National Museum of American History Archives Center, 28, https://sova.si.edu/record/NMAH.AC.1498.

16 Nathan Ensmenger, *The Computer Boys Take Over: Computers, Programmers, and the Politics of Technical Expertise* (Cambridge: MIT Press, 2010).

17 Gita Jackson, "Spacewar! Creators Didn't Know They Were Making History," Kotaku.com (December 5, 2018), https://kotaku.com/spacewar-creators-didnt-know-they-were-making-history-1830887504.

Chapter 5 / BASIC and the Illusion of Coding Empowerment

1 Dartmouth maintains that they were an undergraduate and Kemeny. My research demonstrates that Busch and McGeachie were more likely. For more on the Dartmouth Time-Sharing System and BASIC, see Joy Lisi Rankin, *A People's History of Computing in the United States* (Cambridge, MA: Harvard University Press, 2018).

2 Margot Lee Shetterly, *Hidden Figures* (New York: HarperCollins, 2016), xvi.

3 Arvid Nelsen, "Race and Computing: The Problem of Sources, the Potential of Prosopography, and the Lesson of *Ebony* Magazine," *IEEE Annals of the History of Computing* 39, no. 1 (2016): 29–51.

4 Mar Hicks, *Programmed Inequality: How Britain Discarded Women Technologists and Lost Its Edge in Computing* (Cambridge, MA: MIT Press, 2017); Janet Abbate, *Recoding Gender: Women's Changing Participation in Computing* (Cambridge, MA: MIT Press, 2012); Nathan Ensmenger, *The Computer Boys Take Over: Computers, Programmers, and the Politics of Technical Expertise* (Cambridge, MA: MIT Press, 2010).

5 Joy Lisi Rankin, "For 50 Years, Tech Companies Have Tried to Increase Diversity by Fixing People Instead of the System," *Slate* (March 31, 2021), https://slate.com/technology/2021/03/google-acm-digital-skills-training-diversity-history.html.

Chapter 6 / The First Email

1 Pat Crisman, Glenda Schroeder, Louis Ponzin, "Minimum System Documentation," Programming Staff Note 39, Compatible Time-Sharing System (CTSS), MIT, December 1964 or January 1965, https://multicians.org/thvv/psn-39.pdf.

2 The theory of time-sharing already had come up in the context of the first computer network, the US military's Semi-Automatic Ground Environment (SAGE) system. McCarthy, ever modest, thus was reluctant to take credit for coining the term. See John McCarthy, "Reminiscences on the Theory of Time-Sharing," 1983,

http://jmc.stanford.edu/computing-science/timesharing.html. Dartmouth and Cambridge University also launched early academic time-sharing networks.

3 Tom Van Vleck, "The IBM 7094 and CTSS at MIT," in Compatible Time-Sharing System: Fiftieth Year Overview, IEEE Computer Society, 2003, https://multicians.org/thvv/ah-9-05.pdf.

4 CTSS Programmer's Guide, Section AH.9.05, February 14, 1966, https://multicians.org/thvv/ah-9-05.pdf.

5 J.C.R. Licklider and Albert Vezza, "Applications of Information Networks," *Proceedings of the IEEE* 66, no. 11 (November 1978), 1330–46.

6 William Grimes, "Ray Tomlinson, Who Put the @ Symbol in Email, Dies at 74," *New York Times*, March 7, 2016, B14.

7 Cade Metz, "How the Queen of England Beat Everyone to the Internet," *WIRED*, December 25, 2012, https://www.wired.com/2012/12/queen-and-the-internet/.

8 One example of an RFC is J. White, "A Proposed Mail Protocol," Network Working Group, June 1973, https://tools.ietf.org/html/rfc524; more about the collaborative culture is found in Dave Crocker, "A History of Email," *Washington Post*, March 20, 2012; Brad Templeton, "Reaction to the DEC Spam of 1978," https://www.templetons.com/brad/spamreact.html; Leonard J. Shustek, "The Eudora Email Client Source Code," Computer History Museum, May 22, 2018, https://computerhistory.org/blog/the-eudora-email-client-source-code/.

9 "About CompuServe," https://www.compuserve.com/home/about.jsp; Shustek, "Eudora Email Client Source Code"; Samuel Gibbs, "How did Email Grow from Messages between Academics to a Global Epidemic?," *Guardian*, March 7, 2016, https://www.theguardian.com/technology/2016/mar/07/email-ray-tomlinson-history.

10 Jo Thomas, "Satisfaction in Job Well-Done is Only Reward for E-mail Software Inventor," *New York Times*, January 21, 1997, A10; Paul Keegan, "The Office that Ozzie Built," *New York Times*, October 22, 1995, M49. On the dot-com boom and the broader adoption of Email in the 1990s, see Margaret O'Mara, *The Code: Silicon Valley and the Making of America* (New York: Penguin Press, 2019).

Chapter 7 / The Police Beat Algorithm

1 *The Challenge of Crime in a Free Society: A Report by the President's Commission on Law Enforcement and Administration of Justice*. (Washington, DC: United States Printing Office, February 1967).

2 Saul I. Gass, "On the Division of Police Districts into Patrol Beats," in *ACM '68: Proceedings of the 1968 23rd ACM National Conference* (ACM, January 1968): 459–73, https://doi.org/10.1145/800186.810609.

3 "Computer Aids Kansas City Police Setup: Warnings Relayed." *Christian Science Monitor*, August 26, 1968; "Kansas City Police Unveil Alert System," *Los Angeles Times*, July 11, 1968, 4; IBM, "Catching the Bad Guys," IBM Archives Value One Exhibit, https://www.ibm.com/ibm/history/exhibits/valueone/valueone_bad.html. The following events are reported in "Computers Play a Deadly Game: Cops and Robbers," Think, May 1971, 28–30.

4 Charlton D. McIlwain, *Black Software: The Internet & Racial Justice, from the Afronet to Black Lives Matter* (New York: Oxford University Press, 2019), 234–35, 244.

Chapter 8 / "Apollo 11, Do Bailout"

1 John F. Kennedy, "Address at Rice University on the Nation's Space Effort," September 12, 1962, Rice University, Houston, Texas, transcript and video, https://www.jfklibrary.org/learn/about-jfk/historic-speeches/address-at-rice-university-on-the-nations-space-effort.

2 David A. Mindell, *Digital Apollo: Human and Machine in Space Flight* (Cambridge, MA: MIT Press, 2008).

3 Mindell, *Digital Apollo.*

4 Mindell, *Digital Apollo.*

5 David Scott, "The Apollo Guidance Computer: A User's View," http://klabs.org/history/history_docs/ech/agc_scott.pdf.

6 James E. Tomayko, *Computers in Spaceflight: The NASA Experience* (Washington, D.C.: National Aeronautics and Space Administration, Scientific and Technical Information Division, contractor report, 1988), https://history.nasa.gov/computers/Ch2-5.html.

7 Tomayko, *Computers in Spaceflight.*

8 Jonathan Fildes, "Weaving the Way to the Moon," BBC News, July 15, 2009, http://news.bbc.co.uk/1/hi/technology/8148730.stm.

9 Fildes, "Weaving the Way to the Moon."

10 Dexter Thomas, "The Apollo 11 Computer Code Had References to the Black Power Movement and Quoted Shakespeare," *Los Angeles Times*, July 12, 2016, https://www.latimes.com/science/sciencenow/la-sci-sn-apollo-11-computer-code-20160712-snap-story.html.

11 Jay Bennett, "The (Surprisingly Funny) Code for the Apollo Moon Landings Is Now on GitHub," *Popular Mechanics*, July 11, 2016, https://www.popularmechanics.com/space/moon-mars/a21771/code-for-apollo-moon-landings-on-github/.

12 Don Eyles, "Landing Apollo via Cambridge," *Boston Globe*, 1989, reprinted by *MIT News*, https://news.mit.edu/2009/apollo-eyles-0717.

13 Stephen Witt, "Apollo 11: Mission Out of Control," *Wired*, June 24, 2019, https://www.wired.com/story/apollo-11-mission-out-of-control/.

14 Kennedy, "Address at Rice University on Space Exploration."

Chapter 9 / The Most Famous Comment in Unix History

1 The number of lines in Windows XP was revealed in a January 11, 2011, Facebook post by the official Windows account, https://www.facebook.com/windows/posts/155741344475532.

2 David Cassel, "'You Are Not Expected to Understand This': An Explainer on Unix's Most Notorious Code Comment," New Stack, January 15, 2017, https://thenewstack.io/not-expected-understand-explainer/.

3 Eric S. Raymond, ed., The New Hacker's Dictionary (Cambridge, MA: MIT Press, 1991), available online at http://catb.org/jargon/html/L/Lions-Book.html.
4 Christopher M. Kelty, *Two Bits: The Cultural Significance of Free Software* (Duke University Press, 2008).
5 Dennis Ritchie, "Odd Comments and Strange Doings in Unix," 2004. Ritchie's original web page from Bell-Labs.com has been preserved at archive.org: http://web.archive.org/web/20040206202840/http://cm.bell-labs.com/cm/cs/who/dmr/odd.html.
6 "Computer History Museum Makes Historic MS-DOS and Word for Windows Source Code Available to the Public," press release, March 25, 2014, https://computerhistory.org/press-releases/ms-source-code/?pressalias=ms-source-code.
7 A March 26, 2014, article in *The Verge* titled "See the Hidden Jokes in Microsoft's Early Code" collected up tweets from software architect/developer Leon Zandman sharing his favorite comments. https://www.theverge.com/2014/3/26/5549208/microsoft-early-code-contains-hidden-jokes.
8 Visit that graph online at https://www.vidarholen.net/contents/wordcount/.
9 David Cassel, "Remembering Roger Faulkner: UNIX Champion," New Stack, December 31, 2016, https://thenewstack.io/remembering-roger-faulkner/.

Chapter 10 / The Accidental Felon

1 John Markoff, "Killing the Computer to Save It," *New York Times*, October 30, 2012, https://www.nytimes.com/2012/10/30/science/rethinking-the-computer-at-80.html.
2 Katie Hafner and John Markoff, *Cyberpunk: Outlaws and Hackers on the Computer Frontier* (New York: Simon & Schuster, 1991), 301.
3 John Markoff, "How a Need for Challenge Seduced Computer Expert," *New York Times*, November 6, 1988, https://www.nytimes.com/1988/11/06/us/how-a-need-for-challenge-seduced-computer-expert.html.
4 Hafner and Markoff, *Cyberpunk*, 292.
5 Hafner and Markoff, *Cyberpunk*, 299.
6 Hafner and Markoff, *Cyberpunk*, 304.
7 Daniel Snyder. "The very first viruses: Creeper, Wabbit and Brain," InfoCarnivore.com, May 30, 2010, https://infocarnivore.com/the-very-first-viruses-creeper-wabbit-and-brain/.
8 Markoff, "How a Need for Challenge Seduced Computer Expert."

Chapter 11 / Internet Relay Chat

1 "IRC Is Dead, Long Live IRC," Tech Musings, Solar Winds Pingdom, April 24, 2012, https://web.archive.org/web/20160407082808/http://royal.pingdom.com/2012/04/24/irc-is-dead-long-live-irc/.
2 "Internet Relay Chat," Wikipedia, accessed February 17, 2022, https://en.wikipedia.org/wiki/Internet_Relay_Chat.

3 IRC actions are automatically preceded by an asterisk.
4 These uses of asterisks in non-action messages visually set off performative expressions from other text, while recalling their origins as action commands (see note 3).
5 The use of the third-person subject in virtual performatives likely derived from the practice in chat platforms such as IRC and multi-user dungeons of automatically preceding each utterance with the user's "nick" or character name.
6 "Whacking with a Wet Trout," Wikipedia, accessed February 17, 2022, https://en.wikipedia.org/wiki/Wikipedia:Whacking_with_a_wet_trout.

Chapter 12 / Hyperlink

1 Vannevar Bush, "As We May Think," *Atlantic*, July 1945.
2 Ben Shneiderman and Catherine Plaisant, "Hypertext Research: The Development of HyperTIES," HCIL Archive, University of Maryland Human Computer Interaction Lab, updated July 2018, http://www.cs.umd.edu/hcil/hyperties/.
3 Tim Berners-Lee, "Information Management: A Proposal," CERN, March 1989/May 1990, republished by W3 Archive, https://www.w3.org/History/1989/proposal.html.
4 Tim Berners-Lee, "Answers for Young People," W3 Archive, accessed June 4, 2021, https://www.w3.org/People/Berners-Lee/Kids.html.

Chapter 13 / JPEG

1 Joseph Fourier, *The Analytical Theory of Heat*, The University Press, 1878.
2 Nasir Ahmed, "How I Came up with the Discrete Cosine Transform," *Digital Signal Processing*, 1 no. 1 (1991): 4–5.
3 Fourier, *Analytical Theory of Heat*.
4 Hany Farid, *Fake Photos* (Cambridge, MA: MIT Press, 2019).
5 Farid, *Fake Photos*.

Chapter 14 / The Viral Internet Image You've Never Seen

1 Matthew Prince, cofounder and CEO of Cloudflare, in discussion with the author, July 2021.
2 Jason Kint, in discussion with the author, June 2021.
3 Richard M. Smith, "The Web Bug FAQ," last modified November 11, 1999, https://web.archive.org/web/20071016061254/https:/w2.eff.org/Privacy/Marketing/web_bug.html.
4 Katie Moussouris, founder and CEO of Luta Security, in discussion with the author, June 2021.
5 Lukasz Olejnik, in discussion with the author, June 2021.
6 Ashkan Soltani, in discussion with the author, July 2021.

7 Kenneth White, security engineer and independent researcher, in discussion with the author, July 2021.

8 Lily Hay Newman, "The New iOS Update Lets You Stop Ads from Tracking You—So Do It," WIRED.com, April 26, 2021, https://www.wired.com/story/ios-app-tracking-transparency-advertising/.

9 "Turn images on or off in Gmail," Gmail Help, Google, https://support.google.com/mail/answer/145919.

10 Vinay Goel, "An Updated Timeline for Privacy Sandbox Milestones," The Keyword, Google, June 24, 2021, https://blog.google/products/chrome/updated-timeline-privacy-sandbox-milestones/.

11 Barb Smith, "Developers: Get Ready for New SameSite=None; Secure Cookie Settings," Chromium Blog, Google, October 23, 2019, https://blog.chromium.org/2019/10/developers-get-ready-for-new.html.

12 Vinay Goel, "Get to Know the New Topics API for Privacy Sandbox," *Keyword*, Google, January 25, 2022, https://blog.google/products/chrome/get-know-new-topics-api-privacy-sandbox/.

13 Anthony Chavez, "Introducing the Privacy Sandbox on Android," *Keyword*, Google, February 16, 2022, https://blog.google/products/android/introducing-privacy-sandbox-android/.

Chapter 15 / The Pop-Up Ad

1 Ethan Zuckerman, "The Internet's Original Sin," *Atlantic*, August 14, 2014, https://www.theatlantic.com/technology/archive/2014/08/advertising-is-the-internets-original-sin/376041/.

2 Adrienne LaFrance, "The First Pop-Up Ad," *Atlantic*, August 14, 2014, https://www.theatlantic.com/technology/archive/2014/08/the-first-pop-up-ad/376053/.

3 Shoshana Zuboff, "Big Other: Surveillance Capitalism and the Prospects of an Information Civilization," *Journal of Information Technology* 30, no. 1 (2015): 75–89.

4 Ethan Zuckerman, "The Case for Digital Public Infrastructure," Knight First Amendment Institute at Columbia University, January 17, 2020, https://knightcolumbia.org/content/the-case-for-digital-public-infrastructure.

Chapter 16 / Wear This Code, Go to Jail

1 Adam Back, "Munitions T-shirt," accessed March 8, 2021, http://www.cypherspace.org/adam/uk-shirt.html.

2 United States Munitions List, 22 CFR § 121.1(a).

3 United States Commerce Control List, 15 C.F.R pt. 774 supp 1.

4 David E. Sanger, "High-Tech Exports Hit Antiquated Speed Bumps," *New York Times*, June 13, 1999.

5 David Kahn, *The Codebreakers*, rev. ed. (New York: Scribner, 1996).

6 Steven Levy, *Crypto: How the Code Rebels Beat the Government—Saving Privacy in the Digital Age* (New York: Viking, 2001).

7 Bernstein v. US Dept. of Justice, 176 F.3d 1132 (9th Cir. 1999), *rehearing en banc granted*, 192 F.3d 1308 (Mem), *dismissed as moot sub nomine* Bernstein v. Dept. of Commerce, No. 95–0582 (N.D. Cal. July 28, 2003).
8 Junger v. Daley, 209 F.3d 481 (6th Cir. 2000).
9 Steven Levy, "Battle of the Clipper Chip," *New York Times*, June 12, 1994.
10 Apple, "Software License Agreement for macOS Big Sur," accessed March 8, 2021, https://www.apple.com/legal/sla/docs/macOSBigSur.pdf.
11 David S. Touretzky, "Gallery of CSS Descramblers," last modified Feb. 13, 2008, http://www.cs.cmu.edu/~dst/DeCSS/Gallery/.
12 Universal City Studios, Inc. v. Corley, 273 F. 3d 429 (2d Cir. 2001).
13 Defense Distributed v. US Dept. of State, 838 F. 3d 451 (5th Cir. 2016).
14 Crunchbase, "Blockstream," accessed March 8, 2021, https://www.crunchbase.com/organization/blockstream.

Chapter 17 / Needles in the World's Biggest Haystack

1 Sergey Brin and Lawrence Page, "The Anatomy of a Large-Scale Hypertextual Web Search Engine," *Computer Networks and ISDN Systems* 30, no. 1 (1998): 107–17.

Chapter 18 / A Failure to Interoperate

1 James Oberg, "Why the Mars Probe Went Off Course," *IEEE Spectrum* 36, no. 12 (December 1999): 38; Edward A. Euler et al., "The Failures of the Mars Climate Orbiter and Mars Polar Lander: A Perspective from the People Involved," in *Guidance and Control: 24th Annual AAS Guidance and Control Conference*, ed. Robert D. Culp and Charles N. Schira, vol. 107 (American Astronautical Society: Univelt, 2001), 11; Mars Climate Orbiter Mishap Investigation Board, "Phase I Report," November 10, 1999, 13–14, https://llis.nasa.gov/llis_lib/pdf/1009464main1_0641-mr.pdf.
2 Mars Climate Orbiter Mishap Investigation Board, "Phase I Report," 16; Mars Program Independent Assessment Team, "Mars Program Independent Assessment Team Report," March 14, 2000, 18, https://ntrs.nasa.gov/citations/20000032458; Jet Propulsion Laboratory Special Review Board, "Report on the Loss of the Mars Climate Orbiter Mission," November 11, 1999, 5–6, https://trs.jpl.nasa.gov/handle/2014/38186; Euler et al., "Failures of Mars Climate Orbiter," 6–7; Oberg, "Mars Probe," 36–37.
3 "According to a JPL spokesman, every maneuver intended to dump momentum added a velocity error of about 0.001 meter per second, on a probe that was traveling at a rate of tens of kilometers per second" (Oberg, "Mars Probe," 36).
4 JPL Special Review Board, "Report on Loss," 32–33.
5 Euler et al., "Failures of Mars Climate Orbiter," 6; JPL Special Review Board, "Report on Loss," 36.
6 Mars Climate Orbiter Mishap Investigation Board, "Phase I Report," 16–17; see also Euler et al., "Failures of Mars Climate Orbiter," 6.
7 Mars Climate Orbiter Mishap Investigation Board, "Phase I Report," 16.
8 JPL Special Review Board, "Report on Loss," 1.

9 NASA, "Mars Climate Orbiter Arrival: Press Kit," September 1999, 6, https://www.jpl.nasa.gov/news/press_kits/mcoarrivehq.pdf.
10 JPL Special Review Board, "Report on Loss," 1.
11 JPL Special Review Board, "Report on Loss," 37; *Angular Momentum Desaturation File: Software Interface Specification*, by John E. Ekelund, (NASA, 1995), secs. 4.2.2.6–17, https://pds.nasa.gov/data/mgs-m-rss-1-map-v1.0/mors_0552/document/amd_sis.htm.
12 Euler et al., "Failures of Mars Climate Orbiter," 7.
13 John Palfrey and Urs Gasser, *Interop: The Promise and Perils of Highly Interconnected Systems* (New York: Basic Books, 2012).
14 Palfrey and Gasser, *Interop*, 21–22.
15 JPL Special Review Board, "Report on Loss," 40; Mars Climate Orbiter Mishap Investigation Board, "Phase I Report," 18.
16 Miriam Reisman, "EHRs: The Challenge of Making Electronic Data Usable and Interoperable," *Pharmacy and Therapeutics* 42 (September 2017): 573, https://www.ncbi.nlm.nih.gov/pmc/articles/PMC5565131/; Chinmayi Sharma, "Concentrated Digital Markets, Restrictive APIs, and the Fight for Internet Interoperability," *University of Memphis Law Review* 50 (2019): 455–61, https://www.memphis.edu/law/documents/04_sharma.pdf.
17 Marcus Wohlsen, "Why Copyrighted Coffee May Cripple the Internet of Things," *Wired*, March 6, 2014, https://www.wired.com/2014/03/copyrighted-coffee-undermine-whole-internet-things/.

Chapter 19 / The Code That Launched a Million Cat Videos

1 Lowen Liu interview with Colin Angle, December 6, 2018.
2 "Dave Chappelle Pepsi Commercial," posted on DailyMotion, Paris, France, circa 2008; Bob Garfield, "Pepsi Finally Acknowledges Real Point of Cola in New Ads," *Ad Age*, December 1, 2003.
3 "iRobot Reports Fourth-Quarter and Full-Year 2020 Financial Results," iRobot (press release), PR Newswire, February 10, 2021.
4 Hope Reese, "Joe Jones: Roomba Inventor. Roboticist. Vindicated Pioneer," *TechRepublic*, November 12, 2015.
5 Liu interview with Colin Angle, 2018. See also the interview of Colin Angle in *New York* magazine by Jake Swearingen ("How the Roomba Company Decided to Stop Making Bots for the Military," *New York*, December 6, 2018, https://nymag.com/intelligencer/2018/12/why-the-roomba-company-stopped-making-bots-for-the-military.html).
6 Liu interview with Colin Angle, 2018.
7 Evan Ackerman, "Roomba Inventor Joe Jones on His New Weed-Killing Robot, and What's So Hard about Consumer Robotics," *IEEE Spectrum*, July 6, 2017.
8 Mark J. Chiappetta and Joseph L. Jones, "Navigational Control System for a Robotic Device," US Patent No. US8718821B2, filed September 13, 2002.
9 Ackerman, "Roomba Inventor Joe Jones."
10 Liu interview with Colin Angle, 2018.
11 Reese, "Joe Jones: Roomba Inventor."

Chapter 20 / Nakamoto's Prophecy

1. Satoshi Nakamoto, "Bitcoin: A Peer-to-Peer Electronic Cash System," bitcoin.org, October 31, 2008, http://www.bitcoin.org/bitcoin.pdf.
2. Nakamoto, "Bitcoin," 2008.
3. Georg Simmel, *The Philosophy of Money*, 3rd enlarged edition, ed. David Frisby, trans. Tom Bottomore (1900; New York: Routledge, 2004).
4. Max Weber, *General Economic History*, Social Science Classics Series (1923; New Brunswick, NJ: Transaction Books, 1981).

Chapter 21 / The Curse of the Awesome Button

1. Pete Cashmore, "Facebook's Massive Growth: Hits 30 Million Users," Mashable, January 29, 2009, https://mashable.com/archive/facebook-users-2.
2. Leah Pearlman, interview with the author, May 2021.
3. Victor Luckerson, "The Rise of the Like Economy," *Ringer*, February 15, 2017, https://www.theringer.com/2017/2/15/16038024/how-the-like-button-took-over-the-internet-ebe778be2459.
4. Andrew Bosworth, "What's the History of the 'Awesome Button' (That Eventually Became the Like Button) on Facebook?," Quora, last updated October 16, 2014, https://www.quora.com/Whats-the-history-of-the-Awesome-Button-that-eventually-became-the-Like-button-on-Facebook.
5. Bosworth, "What's the History of the 'Awesome Button.'"
6. Leah Pearlman, interview with the author, May 2021.
7. John Patrick Pullen, "How Vimeo Became Hipster YouTube," *Fortune*, June 16, 2021, https://fortune.com/2011/02/23/how-vimeo-became-hipster-youtube/.
8. Bret Taylor, "I Like It, I Like It," Friendblog (FriendFeed, October 30, 2007), http://blog.friendfeed.com/2007/10/i-like-it-i-like-it.html.
9. Bosworth, "What's the History of the 'Awesome Button.'"
10. Ezra Callahan, interview with the author, May 2021.
11. Leah Pearlman, interview with the author, May 2021.
12. Kate Conger, "Google Removes 'Don't Be Evil' Clause from Its Code of Conduct," *Gizmodo*, May 18, 2018, https://gizmodo.com/google-removes-nearly-all-mentions-of-dont-be-evil-from-1826153393.
13. Reema Khrais, "The Most Powerful Tool in Social Media," *Marketplace*, American Public Media, April 29, 2019, https://www.marketplace.org/2018/04/11/it-was-known-button-it-was-awesome-button/.
14. Bosworth, "What's the History of the 'Awesome Button.'"
15. Tom Whitnah, October 5, 2010, comment on Bosworth, "History of the 'Awesome Button.'"
16. Leah Pearlman, "'I Like This,'" Facebook, February 9, 2009, https://www.facebook.com/notes/10160195053101729/.
17. Chris Crum, "Now You Can 'Like' Comments on Facebook," *WebProNews*, June 16, 2010, https://www.webpronews.com/now-you-can-like-comments-on-facebook/.

18 Jason Kincaid, "50,000 Websites Have Already Integrated Facebook's New Social Plugins," *TechCrunch*, April 29, 2010, https://techcrunch.com/2010/04/28/50000-websites-have-already-integrated-facebooks-new-social-plugins/.

19 Tom Simonite, "Facebook's Like Buttons Will Soon Track Your Web Browsing to Target Ads," *MIT Technology Review*, September 16, 2015, https://www.technologyreview.com/2015/09/16/166222/facebooks-like-buttons-will-soon-track-your-web-browsing-to-target-ads/.

20 "Facebook Reports First Quarter 2021 Results," Facebook Investor Relations, April 28, 2021, https://investor.fb.com/investor-news/press-release-details/2021/Facebook-Reports-First-Quarter-2021-Results/default.aspx.

21 Robinson Meyer and Alexis Madrigal, "How Facebook's Chaotic Push into Video Cost Hundreds of Journalists Their Jobs," *Atlantic*, October 25, 2018, https://www.theatlantic.com/technology/archive/2018/10/facebook-driven-video-push-may-have-cost-483-journalists-their-jobs/573403/.

22 Lucia Moses, "LittleThings Shuts Down, a Casualty of Facebook News Feed Change," *Digiday*, February 28, 2018, https://digiday.com/media/littlethings-shuts-casualty-facebook-news-feed-change/.

23 Pat Garofolo, "Close to Home: How the Power of Facebook and Google Affects Local Communities," American Economic Liberties Project, August 2020, https://www.economicliberties.us/wp-content/uploads/2020/08/Working-Paper-Series-on-Corporate-Power_6.pdf.

24 Tom Warren, "You Can Now Hide Like Counts on Instagram and Facebook," *Verge*, May 26, 2021, https://www.theverge.com/2021/5/26/22454428/facebook-instagram-hide-like-counts-feature-now-available.

25 Leah Pearlman, interview with the author, May 2021.

26 Ezra Callahan, interview with the author, May 2021.

Chapter 22 / The Bug No One Was Responsible For—Until Everyone Was

1 Colm MacCárthaigh, Twitter post, April 7, 2019, https://twitter.com/colmmacc/status/1114947268040327169?s=20.

2 Dan Goodin, "Vicious Heartbleed Bug Bites Millions of Android Phones, Other Devices," *Ars Technica*, April 14, 2014, https://arstechnica.com/information-technology/2014/04/vicious-heartbleed-bug-bites-millions-of-android-phones-other-devices/.

3 Nicole Perlroth, "Companies Back Initiative to Support OpenSSL and Other Open-Source Projects," *New York Times*, April 24, 2014, Technology, https://bits.blogs.nytimes.com/2014/04/24/companies-back-initiative-to-support-openssl-and-other-open-source-projects/.

4 Jon Brodkin, "Tech Giants, Chastened by Heartbleed, Finally Agree to Fund OpenSSL," *Ars Technica*, April 24, 2014, https://arstechnica.com/information-technology/2014/04/tech-giants-chastened-by-heartbleed-finally-agree-to-fund-openssl/.

Chapter 23 / The Volkswagen Emissions Scandal

1 Jack Ewing, *Faster, Higher, Farther: The Inside Story of the Volkswagen Scandal* (New York: W. W. Norton, 2017), e-book, p. 310, chap. "Exposure."
2 Jennifer Chu, "Study: Volkswagen's Excess Emissions Will Lead to 1,200 Premature Deaths in Europe," *MIT News*, March 3, 2017, https://news.mit.edu/2017/volkswagen-emissions-premature-deaths-europe-0303#:~:text=The%20researchers%20estimate%20that%201%2C200,affected%20cars%20sold%20in%20Germany.
3 Lee Vinsel, *Moving Violations: Automobiles, Experts, and Regulations in the United States* (Baltimore: Johns Hopkins University Press, 2019), 176–77; Scott H. Dewey, "'The Antitrust Case of the Century': Kenneth F. Hahn and the Fight Against Smog," *Southern California Quarterly*, 81, no.3 (Fall 1999): 341–76.
4 Vinsel, *Moving Violations*, 189–90.
5 Warren Brown and Pierre Thomas, "Emissions Case Costs GM Fine, Cadillac Recall," *Washington Post*, December 1, 1995.
6 Ewing, *Faster, Higher, Farther*, chap. 6.
7 M. Contag et al., "How They Did It: An Analysis of Emission Defeat Devices in Modern Automobiles," *2017 IEEE Symposium on Security and Privacy (SP)*, 2017: 231.
8 M. C. Forelle, in conversation with author, April 1, 2021.

Chapter 24 / The Code That Brought a Language Online

1 "Shahbag Protests," Wikipedia, revised October 23, 2021, https://en.wikipedia.org/wiki/2013_Shahbag_protests.
2 "Ahmed Rajib Haider," Wikipedia, revised November 23, 2021, https://en.wikipedia.org/wiki/Ahmed_Rajib_Haider.
3 Rohit Khanna, "'Atheist' Tag Used to Attack Bloggers," *Times of India*, June 10, 2015, https://timesofindia.indiatimes.com/india/atheist-tag-used-to-attack-bloggers/articleshow/47606967.cms.
4 Arafatul Islam, "Bangladesh Gags Award-Winning Blogger," *Deutsche Welle*, March 25, 2013, https://www.dw.com/en/bangladesh-gags-award-winning-blogger/a-16697713.
5 "Hefazat Enforces Dawn-to-Dusk Hartal for Monday," RisingBD.com, April 7, 2013, https://www.risingbd.com/english/Hefazat_enforces_dawn-to-dusk_hartal_for_Monday/1536.
6 Joshua Hammer, "The Imperiled Bloggers of Bangladesh," *New York Times*, December 29, 2015, https://www.nytimes.com/2016/01/03/magazine/the-price-of-secularism-in-bangladesh.html.
7 Jana Syeda Gulshan Ferdous, "BTRC Attempts to Silence the Voice of Bloggers," OurTimeBD.com, February 22, 2019, https://www.ourtimebd.com/beta/btrc-attempts-to-silence-the-voice-of-bloggers/.
8 Arild Klokkerhaug, "Transparency Report," SomewhereinBlog.net, February 29, 2013, https://www.somewhereinblog.net/transparency_report#.

Chapter 25 / Telegram

1 “Iran: Tightening the Net 2020,” ARTICLE 19, September 16, 2020, https://www.article19.org/ttn-iran-november-shutdown/.

2 Ali Khamenei, “Supreme Leader’s Speech on November 25, 2009,” Khamenei.ir, https://farsi.khamenei.ir/speech-content?id=8430.

3 Shaun Walker, “Founder of Vkontakte Leaves after Dispute with Kremlin-Linked Owners,” *Guardian*, April 2, 2014, http://www.theguardian.com/media/2014/apr/02/founder-pavel-durov-leaves-russian-social-network-site-vkontakte.

4 Catherine Shu, “Meet Telegram, a Secure Messaging App from the Founders of VK, Russia’s Largest Social Network,” *TechCrunch*, October 28, 2013, https://techcrunch.com/2013/10/27/meet-telegram-a-secure-messaging-app-from-the-founders-of-vk-russias-largest-social-network/.

5 Ryan Gallagher, “Data Breach Shows Iranians Use Chat Apps to Spy, Researchers Say,” *Bloomberg*, April 17, 2020, https://www.bloomberg.com/news/articles/2020-04-17/data-breach-shows-iranians-use-chat-apps-to-spy-researchers-say.

6 Martim Lobao, “Telegram v3.2 Brings Channels for Broadcasting Your Messages to the World,” *Android Police* (blog), September 22, 2015, https://www.androidpolice.com/2015/09/22/telegram-v3-2-brings-channels-broadcasting-messages-world/.

7 Farhad Fatemi, “Feasibility Study for the Transfer of Telegram’s Traffic to Iran: The Public Benefit,” *Arvan Cloud* (blog), September 6, 2017, https://www.arvancloud.com/blog/امکان‌سنجی-انتقال-ترافیک-تلگرام-به-ای/.

8 Mahsa Alimardani, “The Chilling Effect of Officials Discussing Telegram’s Imminent Ban in Iran,” *Global Voices Advox* (blog), April 25, 2018, https://advox.globalvoices.org/2018/04/25/the-chilling-effect-of-officials-discussing-telegrams-imminent-ban-in-iran/.

9 Mahsa Alimardani, “Is Telegram’s Compliance with Iran Compromising the Digital Security of Its Users?” *Global Voices Advox* (blog), accessed July 30, 2021, https://advox.globalvoices.org/2015/08/28/is-telegrams-compliance-with-iran-compromising-the-digital-security-of-its-users/.

10 Mahsa Alimardani, “What Telegram Owes Iranians,” *Politico Magazine*, January 1, 2018, https://www.politico.com/magazine/story/2018/01/01/irans-telegram-revolution-216206/.

11 Amir Vahdat, “Telegram Denies Iran’s Claim It Installed Servers There,” Associated Press, July 30, 2017, https://apnews.com/article/c41a64ac397941dfa16f5267960f1258.

12 “Iran: Block on Telegram is a blow to freedom of expression,” ARTICLE19, May 11, 2018, https://www.article19.org/resources/iran-the-block-on-telegram-in-iran-is-a-blow-to-freedom-of-expression/.

13 Alimardani, “What Telegram Owes Iranians.”

14 Lily Hay Newman, “Iran’s Telegram Ban Has Impacted All Corners of the Country,” *Wired*, June 19, 2018, https://www.wired.com/story/iran-telegram-ban/.

15 “73.6% of People over the Age of 18 in the Country Currently Use Social Media / WhatsApp Messenger Ranks First,” Islamic Students Polling Agency (ISPA),

February 22, 2021, accessed July 25, 2021, http://ispa.ir/Default/Details/fa/2282/73.6-درصد-افراد-بالای-18-سال-کشور،-درحال-حاضر-از-رسانه‌های-اجتماعی-استفاده-می-کنند--پیام-رسان-واتس-اپ-در-رتبه-اول.

16 Matt Burgess, “Switched to Telegram? You Need to Know This about Its Encryption,” *Wired UK*, January 27, 2021, https://www.wired.co.uk/article/telegram-encryption-end-to-end-features.

17 Alimardani, “Telegram’s Compliance with Iran.”

18 Finbarr Toesland, “Iran’s Persecuted LGBTQ Community Seeks Refuge Online,” *Overture Global*, accessed July 30, 2021, https://www.overtureglobal.io/story/irans-persecuted-lgbtq-community-seeks-refuge-online.

19 “Sodomites Parading on Telegram in Shadow of Officials’ Negligence,” Gerdab, accessed July 30, 2021, https://gerdab.ir/fa/news/23030/جولان-همجنس-بازان-در-تلگرام-در-سایه-غفلت-مسئولان-تصاویر.

20 ”Six Telegram Admins Arrested in Ardebil over Promoting Immorality,” خبرگزاری تسنیم | Tasnim News Agency, September 14, 2017, https://www.tasnimnews.com/fa/news/1396/06/23/1518311/6-مدیر-کانال-های-تلگرامی-به-اتهام-ترویج-مسائل-غیراخلاق-در-اردبیل-دستگیر-شدند.

21 “Apps, Arrests and Abuse in Egypt, Lebanon and Iran,” ARTICLE 19, February 22, 2018, https://www.article19.org/resources/apps-arrests-abuse-egypt-lebanon-iran/.

22 Gallagher, “Data Breach.”

23 “73.6% of People,” ISPA.

Chapter 26 / Encoding Gender

1 Marie Hicks, “Hacking the Cis-Tem,” *IEEE Annals of the History of Computing* 41, no. 1 (January 1, 2019): 20–33, https://doi.org/10.1109/MAHC.2019.2897667.

2 David L. Johnson, “The Role of the Digital Computer in Mechanical Translation of Languages,” in *Proceedings of the May 6–8, 1958, Western Joint Computer Conference: Contrasts in Computers*, IRE-ACM-AIEE ’58 (Western) (New York: Association for Computing Machinery, 1958), 161–65, https://doi.org/10.1145/1457769.1457815.

3 Os Keyes, “The Misgendering Machines: Trans/HCI Implications of Automatic Gender Recognition,” *Proceedings of the ACM on Human-Computer Interaction* 2, CSCW (November 2018): 1–22, https://doi.org/10.1145/3274357.

4 Michael Hill, “NY Social Service Agency Sued for Not Allowing X Gender Mark,” AP NEWS, March 29, 2021, https://apnews.com/article/new-york-lawsuits-social-services-coronavirus-pandemic-medicaid-3e2f1cbae9d2acf2f3994df5039375b8.

Contributors

Mahsa Alimardani is an Internet researcher focusing on freedom of expression and access to information online in Iran. Her research aims to understand communications ecologies within Iran's information control space.

Elena Botella, a principal at Omidyar Network, is the author of *Delinquent: Inside America's Debt Machine*. Her work on technology and finance has appeared in the *New Republic*, *Slate*, *Forbes*, and elsewhere.

Meredith Broussard is associate professor at the Arthur L. Carter Journalism Institute of New York University and the author of *Artificial Unintelligence: How Computers Misunderstand the World*. Her research focuses on artificial intelligence in investigative reporting, with particular interests in AI ethics and using data analysis for social good.

David Cassel has been writing about technology for more than 25 years, with articles appearing everywhere from CNN and the *Wall Street Journal*'s site to popular tech news sites like Wired, Gizmodo, and Salon. Since 2015 he's also been a contributing writer for The New Stack, covering the culture of technology (and ways it intersects the broader popular culture), while also expanding his interests into computer programming and app development.

Arthur Daemmrich is the director of the Lemelson Center for the Study of Invention and Innovation at the Smithsonian Institution. He has published in the fields of science and technology studies, history of technology, and business policy, with an emphasis on the interplay of innovation and risk-based regulation.

Charles Duan is a postdoctoral associate at Cornell Tech, where he researches technology and intellectual property law. He is also a senior policy fellow with the Program on Information Justice and Intellectual Property at American University Washington College of Law, a senior fellow for technology and innovation policy with the R Street Institute, and the author of *A Five Part Plan for Patent Reform.*

Quinn DuPont has over a decade of experience researching and writing about crypto. His publications include *Cryptocurrencies and Blockchains* (Polity Press, 2019). Previously he was an assistant professor of business at University College Dublin.

Claire L. Evans is a writer and musician based in Los Angeles. She is the singer and coauthor of the Grammy-nominated pop group YACHT and the author of *Broad Band: The Untold Story of the Women Who Made the Internet* (Penguin Random House).

Hany Farid is a professor at the University of California, Berkeley with a joint appointment in Electrical Engineering & Computer Sciences and the School of Information. He is the author of *Fake Photos* and *Photo Forensics.*

James Grimmelmann is a professor at Cornell Law School and Cornell Tech who studies Internet and intellectual property law.

Katie Hafner is a longtime science and technology writer who has published seven books, including *Cyberpunk: Outlaws and Hackers on the Computer Frontier* (with John Markoff) and *Where Wizards Stay Up Late: The Origins of the Internet* (with Matthew Lyon). She is host and co-executive producer of the narrative podcast *Lost Women of Science.*

Susan C. Herring is professor of information science and linguistics and director of the Center for Computer-Mediated

Communication at Indiana University, Bloomington. She is the editor of *Computer-Mediated Communication: Linguistic, Social and Cross-Cultural Perspectives*, *The Multilingual Internet: Language, Culture, and Communication Online* (with B. Danet), and *The Handbook of Pragmatics of Computer-Mediated Conversation* (with D. Stein and T. Virtanen).

Syeda Gulshan Ferdous Jana, a Bangladeshi journalist, is the cofounder and editor of somewhereinblog.net, the world's first Bengali language blogging platform. Her work has had a significant impact on the freedom of expression and social activism throughout Bangladesh and resulted in several national and international awards.

Lowen Liu is an editor at *Slate*, with an interest in the impact of technology on everyday lives.

John MacCormick is professor of computer science at Dickinson College. He is the author of *Nine Algorithms That Changed the Future: The Ingenious Ideas That Drive Today's Computers* and *What Can Be Computed? A Practical Guide to the Theory of Computation.*

Brian McCullough is the host of the Techmeme Ride Home podcast, Silicon Valley's daily news podcast of record. He is the author of *How The Internet Happened: From Netscape to the iPhone.*

Charlton McIlwain is Professor of Media, Culture, and Communication at New York University and author of *Black Software: The Internet & Racial Justice, From the Afronet to Black Lives Matter*.

Lily Hay Newman is a senior writer at *Wired* focused on information security, digital privacy, and hacking. Previously she was a technology reporter at Slate magazine, and her work has also appeared in Gizmodo, Fast Company, *IEEE Spectrum*, and *Popular Mechanics*.

Margaret O'Mara is the Howard & Frances Keller Endowed Professor of History at the University of Washington, where she writes and teaches about the technology industry and its relationship to modern American politics and society. She is the author of *Cities of Knowledge: Cold War Science and the Search for the Next Silicon Valley* and *The Code: Silicon Valley and the Remaking of America*.

Will Oremus is technology news analysis writer for the *Washington Post*, focusing on the ideas, products, and power struggles shaping the digital world. Before joining the *Post* in 2021, he spent eight years writing about tech for *Slate* magazine and two years as a senior writer for the tech and science magazine OneZero.

Nick Partridge is a writer and science communicator at the Smithsonian's National Air and Space Museum where he leads partnerships and special projects, including the national celebration of the 50th anniversary of Apollo 11. He was previously a public radio host at WUGA in Athens, GA.

Benjamin Pope is a Lecturer in Astrophysics and ARC DECRA Fellow at the University of Queensland, in Brisbane, Australia. He applies modern statistical methods to searching for planets around other stars.

Joy Lisi Rankin, PhD, wrote *A People's History of Computing in the United States*. She leads research at AI Now, focusing on gender, race, and power in artificial intelligence and other algorithmic technologies.

Afsaneh Rigot is an analyst, researcher, and advocate covering issues of law, technology, and LGBTQ, refugee, and human rights. She is a senior researcher at ARTICLE 19 focusing on the Middle East and North African (MENA) human rights issues and international corporate responsibility. She is also a Fellow at the Harvard Kennedy School's Technology and Public Purpose Project (TAPP), an Affiliate at the Berkman Klein Centre (BKC) at Harvard, and an advisor at the Cyberlaw Clinic at Harvard.

Ellen R. Stofan, PhD, is the Undersecretary for Research and Science at the Smithsonian Institution where she oversees the Institution's research centers and sciences museums; Smithsonian Libraries and Archives; Office of International Relations; Smithsonian Scholarly Press; and Scientific Diving Program. She previously led the National Air and Space Museum, served as Chief Scientist at NASA, and currently is on the science team of the NASA Dragonfly mission to Saturn's moon Titan.

Ellen Ullman is a former software engineer whose essays and opinion pieces have been widely published. She is the author of the memoir *Close to the Machine*; of two novels, *The Bug* and *By Blood*; and of the nonfiction collection *Life in Code*.

Lee Vinsel is an associate professor of Science, Technology, and Society at Virginia Tech. He is the author of *Moving Violations: Automobiles, Experts, and Regulations in the United States* and coauthor with Andrew L. Russell of *The Innovation Delusion: How Our Obsession with the New Has Disrupted the Work That Matters Most*.

Josephine Wolff is associate professor of cybersecurity policy at the Tufts Fletcher School of Law and Diplomacy. She is the author of *Cyberinsurance Policy: Rethinking Risk in an Age of Ransomware, Computer Fraud, Data Breaches, and Cyberattacks* and *You'll See This Message When It Is Too Late: The Legal and Economic Aftermath of Cybersecurity Breaches*.

Ethan Zuckerman is associate professor of public policy, information and communication at the University of Massachusetts, Amherst. He is the author of *Mistrust: Why Losing Faith In Institutions Provides the Tools to Transform Them* and *Digital Cosmopolitans: Why We Think the Internet Connects Us, Why It Doesn't and How to Rewire It.*

Index

A page number in *italics* refers to an illustration.